Twelve Diseases
THAT CHANGED OUR WORLD

Twelve Diseases
THAT CHANGED OUR WORLD

IRWIN W. SHERMAN

Department of Biology
University of California
Riverside, California

Department of Cell Biology
Institute for Childhood and Neglected Diseases
The Scripps Research Institute
La Jolla, California

ASM PRESS

Washington, DC

Address editorial correspondence to ASM Press, 1752 N St. NW, Washington, DC 20036-2904, USA

Send orders to ASM Press, P.O. Box 605, Herndon, VA 20172, USA
Phone: (800) 546-2416 or (703) 661-1593
Fax: (703) 661-1501
E-mail: books@asmusa.org
Online: estore.asm.org

Library of Congress Cataloging-in-Publication Data

Sherman, Irwin W.
Twelve diseases that changed our world / Irwin W. Sherman.
p. ; cm.
Includes bibliographical references and index.
ISBN 978-1-55581-466-3 (alk. paper)
1. Epidemics—History. 2. Diseases and history. I. Title.
[DNLM: 1. Disease Outbreaks—history. 2. Communicable Disease Control—history. 3. Communicable Diseases—history. 4. History, Early Modern 1451–1600. 5. History, Medieval. 6. History, Modern 1601-. 7. Socioeconomic Factors. WA 11.1 S553t 2007]

RA649.S44 2007
614.4—dc22 2007023680

10 9 8 7 6 5 4 3 2 1

Printed in the United States of America

Cover: Plague in Rome, Jules Elie Delaunay (1828–1891). Courtesy Wellcome Library, London.

Contents

Preface

The literature on the impact of disease on history is large. It chronicles how illness has affected Western civilizations: in the 14th century plague broke the Malthusian stalemate and provided the impetus to restructure European societies; during the past two centuries genetic diseases altered the fates of the British, Spanish, and Russian royal families and contributed to the rise of Lenin, Franco, and Hitler; in the last 100 years we have witnessed how increased opportunities for disease transmission have decimated populations, created panic, and fostered discrimination. We continue to be painfully aware of the power a disease can wield in effecting social and political changes on a grand scale and how it can reveal and exacerbate social tensions. In the past, disease played a role in colonial expansion in the Americas and Africa and, through demographic pressure and starvation, forced a mass migration of the Irish people; tomorrow in different places and in different ways, another disease may do the same.

Historical perspectives of disease can be valuable for a better understanding of how we, and our forebears, survived the onslaught of "plagues" and how we might avoid some of their consequences: confrontations between immigrants and nativists, discrimination against those with different lifestyles, and the social and political disruptions due to incapacitation and death. Of equal value, and much needed, is an examination of the attempts to control disease and how it was possible to improve the public health. In short, this book is about the lessons we have or should have learned from our past encounters with unanticipated outbreaks of disease and how such understanding can be put to use when future outbreaks occur.

The recent SARS and AIDS pandemics clearly show that our lives, as well as the political and economic fortunes of the developed world and emerging nations, can be influenced by the appearance of a contagious disease. In 2004, alarm bells went off as avian influenza spread across the globe, killing millions of domestic fowl and 113 people. The public asked what measures would be needed to stop its spread so that another 1918 to 1920 flu pandemic, which killed tens of millions of people, would not occur. In 2006 cholera swept through West Africa, striking 20,000 people, and in the United States mumps—no longer thought to be a threat because of childhood vaccination—broke out in Iowa and quickly spread to neighboring states, affecting 1,000 people.

These unanticipated epidemics provoke questions. What is needed to curtail the transmission of a disease? What will it take to contain a disease so that protective measures can be instituted? These questions, perplexing and complex, need answers. To simply catalog past diseases and tell of their historic consequences would not be of lasting value to the general public. Rather, it was my feeling that the answers to how we might deal with "coming plagues" could be better obtained by an examination of how past encounters with disease allowed for better control and improved health.

Our world has experienced so many diseases that it would be pointless to deal with all of them. In fact, it would be a nearly impossible task, and, if achieved, it would be numbing to read. Instead, I have selected a dozen diseases that have shaped our history and illuminated the paths taken in finding measures to control them. Porphyria and hemophilia (chapter 1) influenced the political fortunes of England, Spain, Germany, Russia, and the United States; late blight (chapter 2) spawned a wave of immigration that changed the politics of the United States; cholera (chapter 3) stimulated sanitary measures, promoted nursing, and led to the discovery of oral rehydration therapy; smallpox (chapter 4) led to a vaccine that ultimately eradicated the disease; plague (chapter 5) promoted quarantine measures and attenuated vaccines were the result of outbreaks of tuberculosis (chapter 6); syphilis (chapter 7) provided the impetus for cure through chemotherapy; and malaria and yellow fever (chapters 8 and 9) provided the basis for vector control. However, despite these successes, two pandemics—influenza (chapter 10) and HIV/AIDS (chapter 11)—continue to elude control. In this book I try to answer why this is so.

The message of this book is simple: understanding past outbreaks of disease can better prepare us for those in our future. The twelve diseases chosen have influenced the way we look at sickness and show how they resulted in public health measures and other interventions to stem the spread of that disease and others. To eliminate the fear and confusion surrounding "coming plagues," I describe the ways we have succeeded in bringing certain diseases under control and, in other cases, our failures. My purpose in writing this book for the general reader is to show that despite the challenges which an unanticipated illness may place before us, the future is not without hope or remedy.

1

The Legacy of Disease: Porphyria and Hemophilia

In 1962 the U.S. President John F. Kennedy said, "Life is unfair. Some people are sick and others are well." He, of course, was referring to himself and the persistent rumors about his ill health. Forty years later, an examination of his medical records revealed that he had Addison's disease, a life-threatening lack of adrenal gland function, as well as osteoporosis and persistent digestive problems. He was given pain killers (demerol and methadone), stimulants, and antianxiety agents, as well as hormones (hydrocortisone and testosterone) to keep him alive, especially during times of stress. Although doubts linger whether President Kennedy's physical ailments influenced the manner by which the Cuban missile crisis was handled or whether they affected other political decisions, it is clear that for many world leaders, including Great Britain's King George III, several of Queen Victoria's children and grandchildren, Tsar Nicholas II of Russia, and Alfonso XIII and Generalissimo Franco in Spain, as well as, indirectly, the leaders of Nazi Germany, sickness was the seed for historical change.

Porphyria

Madness in the monarchy

Mary Queen of Scots (1542 to 1587) had a mysterious ailment. At the age of 24 she wrote, "Oftentimes I have great pains . . . ascending unto my head . . . it descends to my stomach so that it makes me lack an appetite . . . and there is sickness with great vomit . . . excuse my writing, caused by the weakness of my arm . . . wherewith we are tormented." In 1570, when she had another attack, her symptoms were described by her physician: "terrible pains in the side made worse by every movement, even breathing. She vomited continuously, more than 60 times, and eventually brought up blood. She became delirious, and two days later she lost her sight and speech, had a series of fits, remained unconscious for some hours and was thought to be dead. Yet within 10 days she was up and about again. She had unquiet and melancholy fits, convulsions, shivering, difficulty in swallowing, altered voice, weakness of arms and legs so that she could neither write, walk or even stand unaided." The onset of her symptoms was rapid and suggested to some in her court that she was being poisoned. Others judged her to be hysterical. It is most likely, however, that Mary Queen of Scots was neither hysterical nor the victim of poisoning. Instead, she and many of her descendants probably suffered from an inherited disorder—a curse of British royalty—that would alter the course of world history.

Mary Stuart, Queen of Scotland since birth, was engaged at the age of 3 to Prince Francis, heir to the throne of France; at age 15, when she married him, he was already King Francis II, so she became Queen of France as well as of Scotland. Such glory did not last very long. Francis II died unexpectedly a year after the marriage, and Mary returned to Scotland, where she later married Henry Stuart, Lord Darnley, a relation of the English royal family who was described by historians as a drunkard and an imbecile. Mary did not trust Darnley with affairs of state, and she had several male secretaries who advised her. One of her favorites was David Rizzio, who provoked such jealousy in Mary's husband that he arranged for Rizzio to be murdered. Darnley himself was murdered a year later, and it was widely believed that James Hepburn, 4th Earl of Bothwell, had conspired with the Queen to kill her husband. Shortly thereafter, Mary married Bothwell. This, together with Mary's episodes of blindness, depression, and inability to speak or stand, so disturbed the Scottish nobility that they forced her to abdicate the throne. Mary sought refuge and protection in England, where her cousin Elizabeth I was queen. Mary was an ungrateful and tormented guest in England and became involved in plots to kill Elizabeth. The plots were discovered, and in 1587 Mary was beheaded.

When Elizabeth I died in 1603, Mary's son, James VI of Scotland, succeeded her as James I of England. The King had a disease similar to that of his mother. According to his physician, Sir Theodore de Mayerne, "He was afflicted with pain . . . under his ribs . . . he glows with heat, and his appetite falls off; he sleeps badly; he readily vomits, at times so violently that his face is covered with red spots for two or three days . . . very often he suffered from painful colic . . . with vomiting and diarrhea, preceded by melancholy and nocturnal rigors . . . he had such pain and weakness in the foot that it was left with an odd twist when walking . . . In 1616 . . . for 4 months he had to stay in bed or in a chair . . . In 1619 . . . he sweats easily . . . often suffers bruises . . . he is of exquisite sensitiveness and most impatient of pain . . . He often passed urine red like Alicante wine." Although diagnosis of medical conditions in persons living so long ago is uncertain, it is very likely that the mysterious disease suffered by Mary Queen of Scots, her son King James, and many of their descendants was porphyria, derived from the Greek word "porphuros," meaning "purple," with its telltale sign of red-purple urine.

Gene failure

Porphyria, a hereditary error of metabolism, is linked to the body's production of the pigment hemoglobin, which gives color to our red blood cells and grabs oxygen molecules as blood courses through the lungs. Hemoglobin consists of a protein, globin, coupled to a nonprotein molecule, heme. Heme, an iron complex within a ring structure called porphyrin, is synthesized in the red cells and liver. The reverse of this process, that is, the breakdown of heme to salvage the iron, results in the formation of bile pigments which are stored in the gallbladder and function as a detergent to emulsify fats for easier digestive action; bile pigments also color the feces brown. If there is a block anywhere in the eight-step pathway of heme formation, heme is not produced and the porphyrin intermediates accumulate in a variety of tissues in the body.

The pathway of heme manufacture can be thought of as if it were a river flowing downstream with a series of eight waterwheels along the way; each waterwheel is a cellular factory for making heme intermediates, i.e., porphyrins. To allow for control of water flow, a series of sluice gates are positioned ahead of each waterwheel. For a waterwheel to turn, each sluice gate must be opened by a gatekeeper. When a gatekeeper cannot open a gate, the flow of water is interrupted; water accumulates behind the waterwheel and spills over. Similarly, in the pathway for the synthesis of heme, the gatekeepers are the eight biological catalysts, called enzymes, that allow a controlled flow of intermediates in the pathway. If a gatekeeper "falls asleep at the wheel," i.e., a particular enzyme does not function properly (or is absent), the normal pathway to heme is blocked and porphyrins accumulate in front of the block. These increased amounts of porphyrins do their dirty work, causing abdominal pain and neuropsychiatric symptoms (such as those seen in Mary and James), although the precise molecular basis for this is not known. The porphyrins in the skin, when exposed to UV light, become "excited" and in this state react with molecular oxygen to form activated oxygen, which can lead to cell death, with redness and blistering of the skin and scarring. The porphyrin intermediates also spill over into the urine during an attack. In an individual with porphyria, fresh urine is colorless, but on exposure to air and light for several hours it turns the color of port wine.

How did James I get porphyria from his mother, Queen Mary? The disease was transmitted through inheritance, not by contagion. Porphyria is transmitted as an autosomal dominant; that is, it is not carried on either the X or Y chromosome but on one of the other 44 chromosomes. The presence of a single copy of the defective gene has noticeable effects on the body. If one parent carries the defective gene, then, on average, half the children will bear the defective gene; this gene encodes a defective enzyme and causes porphyria. Most often the disease arises from a partial deficiency in a liver enzyme in the third step (or sometimes the seventh step) of heme synthesis. Porphyria is often more prevalent in females than in males, but in both sexes the symptoms rarely occur before puberty. These days, there are treatments for the disease symptoms, and dietary measures can reduce overproduction of porphyrins; however, in the time of Queen Mary and King James there was no such help.

The curse of the British royal family

The pedigree of porphyria can be traced back as far as Mary Queen of Scots. Her grandson Henry Frederick, Prince of Wales (the eldest son of James I), is believed to have inherited porphyria. His younger brother Charles was next in line for succession to the throne (he reigned as Charles I from 1625 to 1649); he did not have symptoms suggestive of porphyria. Charles I had a daughter named Henrietta Anne, who married a brother of Louis XIV of France and became the Duchess of Orleans. She had symptoms of porphyria and died unexpectedly at the age of 26. The eldest surviving son of Charles I, Charles II, did not appear to have porphyria. However, he had no legitimate children and was succeeded by his younger brother, James II, whose daughter Mary had married William of Orange, ruler of the Netherlands. In 1688, King William was invited to invade England, and James II was forced to leave the country for France. In effect, Mary (the daughter of James II) and William ruled both Britain and the Netherlands. Mary appeared not to suffer from porphyria. She died of smallpox in 1694, leaving William as the sole ruler. After his death in 1702, Mary's younger sister, Anne, became Queen. Queen Anne suffered from indigestion, hysteria, fits of depression, convulsions, and muscular weakness. Indeed, this royal invalid was so weak at age 39 that she had to be carried to her own coronation. She died in a coma at the age of 49, having outlived her longest-surviving child, William Henry, Duke of Gloucester. Her death left the country without a Protestant heir in the direct line of descent. Succession to the throne then passed from the House of Stuart to the House of Hanover through descent from Elizabeth, daughter of James I (and granddaughter of Mary Queen of Scots), who had married the King of Bohemia. Although Elizabeth showed no signs of porphyria, she must have transmitted the defective gene to her daughter Sophia, who married the Elector of Hanover, Ernst August, the father of King George I of Britain. Both George I and George II were healthy, but the grandson of George II, George III, manifested many symptoms of porphyria.

In his play "The Madness of George III," Alan Bennett gives the King a voice:

> "Why do you shiver? I am not cold. I am warm. I am burning. No, I am not burning. It is my body that is burning. And I am locked inside it . . . Well give me my shirt then. What shirt is this? No. It's rough. Feel. It's like calico. Sailcloth. It's a hairshirt . . . These are not my proper stockings. They itch, too. I burn all inward. My limbs are laced with fire. But I will not give into it . . . Oh God, my blood is full of cramps, lobsters crack my bones, there are stones in my belly." Then the King's two servants remark, "Look. What? It's blue. I'd call it purple. You and me, we piss plain. Kings piss purple . . . It has been blue since His Majesty has been ill." The King continues: "Peace of mind! I have no peace of mind. I have had no peace of mind since we lost America . . . All ours. Mine. Gone. A paradise lost. The trumpet of sedition has sounded. We have lost America. Soon we shall lose India, the Indies, Ireland even, our feathers plucked one by one, this island reduced to itself alone, a great state moldered into rottenness and decay. And they will lay it at my door . . . I am not going out of my mind; my mind is going out of me . . . I don't know. I don't know. Madness isn't such a torment. Madness is not half blind. Madmen can stand. They skip. They dance. And I talk. I talk. I hear the words so I have to speak them. I have to empty my head of words. Something has happened. Something is not right. Oh . . . God, please restore me to my senses, or let me die directly for Thy Mercy's sake."

King George III was stubborn and unpredictable in his behavior. Although he did have an attack of ill health in 1765, at age 26, there is no indication that this was accompanied by a fit of madness, and it is not certain whether this disease was porphyria or some other malady. Thus, the American War of Independence in 1775 was precipitated not so much by the King's porphyria as by his inflexible attitude and because he backed the policies of his Prime Minister, Lord North, in the passage of the unpopular Stamp Tax in 1765. "The king's bad judgment may have prevented an amicable settlement, but his faults were shared by his ministers, the majority of the House of Commons and a large proportion of the British public."

Although the loss of the American colonies in the War of Independence between 1775 and 1781 appears not to be attributable to the "royal malady," the same cannot be said for the hostile relations between the Irish and the English. During this period, Protestant settlers and native Catholics lived amicably in Ireland, but in 1798 the Catholics, who were not allowed to sit in the Dublin Parliament, began to revolt, encouraged by France. The uprising was suppressed, and William Pitt, the Prime Minister, suggested that the Dublin Parliament abolish itself and declare union with Britain, with the understanding that Catholics would be eligible to sit in Parliament. In short, Pitt had committed himself (and Britain) to Catholic emancipation. However, Pitt did not inform George III, who regarded himself as "Defender of the Protestants." The King objected to Catholic emancipation, and Pitt resigned. Ten days later, George III suffered an acute attack of what was almost certainly porphyria, and when he recovered a month later, he called back Pitt, who gave the King his solemn promise that Catholic emancipation would never again be mentioned during George's lifetime. This shelved the subject for 28 years, and the Pitt pacification plan for Ireland was doomed to failure. In consequence, the Irish Catholic union with Britain resulted in domination by the alien and oppressive Protestants. Two hundred fifty years later, the troubles in Ireland, due in part to King George's "madness," continue.

King George is thought by many to have had eight porphyric attacks between 1762, when he was 24, and 1804, when he was 66, although it has been argued that his episodes of ill health prior to 1788 did not include symptoms consistent with porphyria. In 1810 he again suffered an attack and lapsed into madness for 2 years; his son, the Prince of Wales, became Regent under the Regency Act of 1811. Because there were hopes that King George might recover, the Prince Regent did not replace his ministers;

however, the King never rebounded sufficiently to resume the throne, and he died blind and deaf at age 81.

In 1968, Ida Macalpine and Richard Hunter, a mother-and-son team of psychiatrists, carefully reviewed the clinical features shown by George III and theorized that his behavioral aberrations were not due to his being "mad" (or, in the modern sense, a manic-depressive psychotic); instead, they concluded that the King's symptoms were consistent with porphyria, a metabolic disorder that caused gastrointestinal symptoms, dermatitis, and dementia. He was, according to Macalpine and Hunter, a simple victim of his "bad," not his "mad," genes. But why the severity and late onset of the attacks of lameness, abdominal and limb pain, racing pulse, insomnia, temporary mental disturbance, and discolored urine? It has been suggested, based on a recent (2005) chemical analysis of a hair obtained from George III, that this was the result of continual exposure to arsenic and/or antimony in the medicine (emetic tartar) commonly prescribed in the 18th century to reduce fever.

George IV, son of George III, also had symptoms consistent with porphyria. While Prince of Wales, he married his cousin Caroline of Brunswick, who was also very probably porphyric. Their only child, Princess Charlotte, also very probably suffered from porphyria. She died in childbirth at the age of 21, possibly partly as a result of this condition. When George IV died, he was succeeded by his brother William Duke of Clarence, who ruled as William IV. Since William had no legitimate children, his heir was his younger brother Edward Duke of Kent, who was most probably porphyric. Edward predeceased the King, and his daughter, Princess Victoria of Kent, became heir. She succeeded her uncle in 1837. Queen Victoria was not porphyric, but she had another hereditary disease, which is discussed later in this chapter. Subsequent British monarchs have shown no signs of porphyria, although in 1968 two living female descendants of the House of Hanover were reported to be porphyric. Thus, although the present House of Windsor appears to be free of porphyria, the disease has persisted in the House of Hanover into the 21st century.

The medical treatment of George III included methods popularly employed to handle the insane: straitjackets, coercion, cupping, and bleeding. However, it is now clear that the "mad King" was misdiagnosed. The illness from which George III presumably suffered, porphyria, and its attendant consequences such as melancholia, depression, sweats, and fits of mania led to the establishment of psychiatry (at that time called the "mad business") as a branch of medicine. The hereditary "madness" due to porphyria has afflicted the British monarchy and the House of Hanover and affected world history for over 500 years!

Hemophilia

Blood will tell

Queen Victoria (born in 1819), who reigned as Queen of the United Kingdom from 1837 to 1901, was in part responsible for bringing the Bolshevik Party into power, contributed to the demise of the House of Romanov, influenced the rise of Generalissimo Franco in Spain, and even arguably played an unwitting role in the ascendancy of the Third Reich in Germany. She did this not through her politics or her armies but through her genes, for she sowed the seeds of a debilitating and potentially fatal disease among the crowned heads of Europe by marrying off her daughters and granddaughters to them, with devastating effects on some of the royal houses concerned.

The disease that Queen Victoria passed on to her offspring was hemophilia or "bleeders' disease." Hemophilia (literally "love of blood") involves a failure of the blood to clot within a normal time. The defect is caused by a missing protein in the plasma, the liquid part of the blood, which is necessary for clot formation. Normal blood may take 5 to 15 min to clot, but in persons with hemophilia (hemophiliacs) the process may take hours or even days. The danger for a person with hemophilia is that even a small wound or bruise may lead to severe and uncontrolled internal bleeding and death.

Without clot formation, the blood flows freely from a wound until the circulatory system collapses—the afflicted person hemorrhages to death. Blood clotting is a complex affair involving a cascade of protein-protein interactions that converts a soluble protein of blood plasma, fibrinogen, into insoluble protein fibers of fibrin. The clotting cascade is like the Mother Goose rhyme "This is the house that Jack built": This is the cat, that killed the rat, that ate the malt, that lay in the house that Jack built. In the clotting cascade: This is the break in the skin, so factor VIII can begin, converting prothrombin to thrombin; when thrombin converts fibrinogen to fibrin, the cross-linked result produces clottin'.

Eighty-five percent of all persons with hemophilia lack factor VIII, one of the clotting factors; the remainder lack factor V. In the absence of such factors, the individual may suffer internal hemorrhages after a minor bump or may die at an early age due to a bleeding crisis. It is possible to diagnose hemophilia as early as the eighth week of pregnancy by DNA hybridization techniques, but in the time of Queen Victoria no such test was available. In recent years, hemophiliacs could be treated with intravenous transfusions of a concentrated form of factor VIII that had been prepared from normal plasma. This form of treatment dramatically lengthened the life expectancy of hemophiliacs from about 20 years to more than 60. However, this therapy was unavailable before 1960, and even when it did become possible to correct hemophilia with transfusion of factor VIII, the dangers of the recipient becoming infected with human immunodeficiency virus and hepatitis virus from such preparations were great indeed. This complication of virus-contaminated preparations has been avoided since 1986, when the gene for factor VIII was cloned, making it possible to synthesize large quantities of "pure" factor VIII in virus-free tissue culture cells.

What is the cause of defective factor VIII? It can be the result of mutations (a change in a single nucleic acid base in the DNA) that produce a shortened version of factor VIII, leading to severe hemophilia, or there can be a complete absence of factor VIII, also leading to severe hemophilia. However, if the mutation results in the insertion of a "wrong" amino acid in factor VIII, the resulting hemophilia is mild.

"Catching" hemophilia

How did Queen Victoria transmit hemophilia to some of her children and grandchildren? Indeed, how did she herself come to be a carrier? Our gender is determined at the moment of fertilization. Each of our somatic (body) cells contains within its nucleus 44 autosomes and one pair of sex chromosomes. During the formation of sperm and eggs, two kinds of sperm are possible (one with an X and one with a Y chromosome) but only one kind of egg occurs (with an X chromosome). Determination of the gender of an offspring depends on the sex chromosome of the fertilizing sperm. If the fertilizing sperm carries an X chromosome, the offspring will be female, and if the fertilizing sperm carries a Y chromosome, the offspring will be male. Genes that are carried on either the X or Y chromosome are called sex-linked genes. The defective gene for hemophilia is carried only on the X chromosome. Since males have only one X chromosome, they have symptomatic hemophilia if they carry the defective form of the gene for factor VIII. However, females, having two X chromosomes, would have to have a double dose of the defective gene to show signs of hemophilia. This is unlikely since the chance of a person having both a hemophiliac father and a carrier mother is quite remote, and females who are hemophiliac die before maturity because the onset of menstruation is fatal.

Hemophiliac fathers pass on the recessive gene to all their daughters but not to any of their sons, because the son receives a Y chromosome, not an X chromosome, from the father. Carrier mothers, i.e., those who carry one normal and one defective gene, have a 50% chance of passing the defective gene to their offspring; affected sons are hemophiliacs, and affected daughters are carriers.

Since there is no record of hemophilia in Queen Victoria's ancestors, it is presumed that either she developed a mutation in the gene for factor VIII in her embryonic cells or a mutation occurred in the X chromosome of one of her parents' germ cells. An alternative possibility, although one for which no real evidence exists, is that she was the illegitimate daughter of a hemophiliac father. Queen Victoria had nine children by her husband, Albert, Prince of Saxe-Coburg and Gotha. Princess Alice (1843 to 1878), Victoria's third child and second daughter, married Prince Louis of Hesse at an early age and had seven children, one of whom, Frederick, was a hemophiliac who died at the age of 3 after falling out of a window. Princess Alice, along with her youngest daughter, May, died of diphtheria in 1878. Her sixth child, Alix, was only 6 years old when her mother died. Alix was a favorite grandchild of Queen Victoria, who hoped that Alix would marry Albert Victor (Duke of Clarence and Avondale), the Queen's grandson and the eldest son of the Prince of Wales (later Edward VII). Alix, however, did not take to the unimpressive Albert Victor, who was rather deaf and somewhat retarded. Had such a union been consummated, Alix's carrier status for hemophilia could have introduced the disease into the British royal family. Instead, she introduced the defective gene into the House of Romanov, the royal family of Russia, and thus contributed to its downfall.

Death of the House of Romanov

Alix first met the Tsarevich Nicholas when she was 12 years of age; 5 years later, they met again and fell in love; they married in 1894, 1 week after the death of Nicholas' father, Tsar Alexander III. Although hemophilia had already been recognized in Victoria's descendants (her son Prince Leopold Duke of Albany died of hemophilia, as did her grandson Frederick of Hesse), the risk was largely unappreciated and/or the value of marrying into a powerful royal house (and a potential ally) took precedence over prudence. On her marriage to the 26-year-old Tsar Nicholas II, Alix took the name Alexandra Feodorovna. Their first four children, born between 1895 and 1901, were girls; this made Alexandra increasingly neurotic since the first duty of a Tsarina was to maintain the House of Romanov by producing a male heir. In 1904, they had a son, Alexis. Alexandra soon discovered that the Tsarevich Alexis was bleeding excessively from the umbilicus and that he had inherited hemophilia. The fragile health of her longed-for son caused her to become more and more withdrawn. She dwelt morbidly on the fact that she had transmitted the disease to her heir. Alexis' condition was kept secret from everyone except close family and their physicians because such a defect would have been regarded as a sign of divine displeasure since the Tsar was both head of the Church and leader of the Russian people. In the summer of 1907, Alexandra was introduced to a "holy man," the monk Gregorii Rasputin (who was born in Siberia in about 1860 to 1865). Rasputin's appearance and demeanor were those of a disheveled vagrant; in addition, he was debauched, alcoholic, coarse, lecherous, and a rapist. He was, however, a charismatic man and a great hypnotist. He recognized and encouraged the Tsarina's fascination for the Russian spirit and her desire to be the soul-mother of its simple people. More importantly, he was able to soothe and calm the distressed and sometimes hysterical Alexis during bouts of hemophilia and hence help stop the bleeding. Increasingly the Tsar and Tsarina came to depend on him. Indeed, in 1907 Alexis recovered from a near-death experience when Rasputin simply stood at the foot of the bed and prayed; he never once touched the child. Again, in 1912, when Alexis was 8 years old, he was bruised while playing in a bathtub and hemorrhaging began. Alexandra once again contacted Rasputin, who responded by telegram that all would be well, and almost miraculously Alexis began to recover. As a result, Rasputin enjoyed increasing personal and political influence with the Tsar and Tsarina, influence which he did not hesitate to take advantage of.

The assassination of Archduke Ferdinand and his wife Sophie in Sarajevo on 28 June 1914 by a Serbian extremist signaled the beginning of the Russian Revolution and the end of the House of Romanov. Three days before Ferdinand, heir to the throne of the Austro-Hungarian Empire, was shot dead, Alexis had slipped on a ladder on his father's yacht, sustaining an injury that resulted in excessive bleeding around the ankles. To complicate matters further, Rasputin had been stabbed in his hometown in Siberia and was unable to minister to the seriously ill Tsarevich. Regarding the international situation, Rasputin wrote from his sick bed: "Let Papa [Nicholas] not plan for war, for with war will come the end of Russia and yourselves and you will lose to the last man." For once, Nicholas ignored Rasputin's advice and mobilized the army against Austria. As a result of the Triple Alliance between Germany, Austria, and Italy, this action meant that Tsar Nicholas of Russia was at war with his cousin-in-law Wilhelm II of Germany, who in turn was at war with his cousin King George V of Great Britain. In the first year, Russia lost 4 million men. After the Tsar took over as Commander-in-Chief in 1916, the results were even more disastrous, and Nicholas was seen as personally responsible. Nicholas' position as Commander-in-Chief took him away from St. Petersburg, and Alexandra was left to govern in his absence. While she ruled the country, Rasputin ruled her. He prevailed upon her to make several government appointments, and the positions were filled by individuals who turned out to be unfit for their duties. The turnover rate among these officials was high, adding instability to incompetence. Both Rasputin and Alexandra were hated by the Russian people, not least because of Alexandra's German origins, which led to accusations that she was a traitor. An increasingly high mortality rate among the soldiers at the front, as well as Alexandra's urging that liberal reforms be abandoned and the Tsar become more autocratic, led to even further hatred of the Tsar by the Russian people. In December 1916, in an attempt to free the Tsar and Tsarina from Rasputin's influence, Prince Youssoupov and Grand Duke Pavlovitch, the Tsar's nephew, assassinated Rasputin. They were punished by being exiled, an action that drove a wedge between Nicholas and the rest of the Romanov family.

Early in 1917, conditions in St. Petersburg deteriorated even further: food and fuel were scarce, people had to queue for hours in the bitter cold to buy bread, and revolution began to brew in the streets. Nicholas ordered, "I command that the disorders in the capital shall be stopped tomorrow as they are inadvisable at the heavy time of war with Germany and Austria." The troops were no longer on his side and did not respond; the soldiers who were garrisoned in St. Petersburg were of no help since they were already consorting with the revolutionaries. The rebellious crowds took over the city, and a provisional government was established. The provisional government attempted to maintain the Romanov dynasty as a constitutional as opposed to an autocratic monarchy by demanding that Nicholas abdicate in favor of the Tsarevich Alexis, with Grand Duke Michael (the Tsar's brother) as Regent; the Army commanders also urged Nicholas to abdicate. Because of his unpopularity and recent ill health, he eventually agreed. However, instead of abdicating in favor of his son, he assigned the throne to his brother Grand Duke Michael and excluded the frail Tsarevich from the succession. The new government, which had been prepared to accept Alexis as a constitutional monarch, feared that Grand Duke Michael might prove to be as autocratic as Tsar Nicholas. Sensing this, Michael abdicated a day after being named Tsar. Anarchy resulted, and the Bolshevik Party, which promised bread, land, and peace, rose to power. As the various political parties struggled for power, the country descended into civil war, and the provisional government feared for the safety of the imperial family. In the spring of 1917, the government approached Great Britain with a request to grant asylum to Nicholas and his family, but the Tsar's cousin, George V, declined. The Tsar and his family were then sent to Tobolsk in Siberia, and in April 1918, after the Bolsheviks had seized power, they were transferred to Yekaterinburg in the Urals. They remained confined at Ipatiev House in Yekaterinburg until the summer of 1918. On 16 July 1918, Nicholas, Alexandra, their five children, their physician, and three servants were taken to the basement, where they were assassinated in a hail of bullets by a Bolshevik firing squad. The Romanov dynasty, which had lasted for over 300 years, had come to an end. It is possible that had the Tsarevich been healthy enough to be named as constitutional monarch after the abdication of his father, the political system might have stabilized and the Bolshevik revolution might have been avoided.

The Spanish royal family bleeds out

The youngest of Queen Victoria's nine children, Beatrice, was born in 1857. She was a hemophilia carrier. She married Henry of Battenberg and transmitted the gene to three of her four children; the eldest son was unaffected. The second son, Leopold, was a hemophiliac. He joined the King's Royal Rifle Corps, but because he was physically delicate and lame, he never saw active service; he died in 1922 following a hip operation. The third son, Maurice, also probably a hemophiliac (although this has been disputed), joined the King's Royal Fusiliers and died of wounds received at the battle of Ypres. The only daughter, Victoria Eugenie (known as Ena), who was a carrier for hemophilia, married Alphonso XIII, the King of Spain; her condition had a significant impact on the political stability of Spain.

The shortage of healthy heirs from the marriage of Eugenie and Alphonso contributed to anti-British feeling in Spain since it was believed that the British had defiled the royal blood of Spain by imposing a genetically defective wife on the Spanish monarch. Unfortunately for Ena, her status as the origin of this disease in the Spanish royal family led to tension in her marriage and its eventual breakdown. Although technically Spain was a constitutional monarchy, in actuality the political parties were weak and so the Spanish King was responsible for appointing his governments. At the end of World War I, the position of the monarchy was further weakened by strikes, assassinations, and a military disaster in Morocco. In 1923, General Miguel Primo de Rivera orchestrated a coup and seized dictatorial power. The King named him Prime Minister, thus appearing to support him. The dictatorship was initially successful and popular, but the people eventually tired of living under a dictator, and Primo de Rivera was unable to sustain his position in the face of economic instability in the late 1920s. The last straw came when the military withdrew support after he imposed some unpopular reforms. He resigned in 1930. The King also lost popular support, and he and his family went into voluntary exile in 1931. Spain became a republic. During the next 5 years, various political groups struggled for power: in 1933 the moderate conservatives were elected, but by 1935 they were replaced by the leftists; military leaders then plotted to overthrow the leftist government. In 1936, Generalissimo Francisco Franco, who had built his reputation in the Moroccan wars of the 1920s, led a Nationalist revolt against the leftists. Franco's justification for his revolt was the defense of Catholic values against all enemies including communism, liberalism, and separatism. The political situation, however, continued to deteriorate, with revolts and murders and finally the Civil War (1936 to 1939). Franco's regime did not tolerate insubordination or political opposition, and those who were opposed to the militaristic society were purged, executed, or imprisoned.

The monarchists in Spain pressed Franco to restore the monarchy. Indeed, monarchism was strong among the generals who had backed him. However, Franco feared that restoration of a liberal constitutional monarchy would be both anti-Catholic and anti-Nationalist. Further, the possible royal heirs were now living in exile. Alphonso, the eldest son, who inherited the hemophilia gene from his great grandmother Queen Victoria, renounced his claim to the throne in order to marry a commoner, although his removal from the succession as a result of his hemophilia was already being considered. He died at age 31 from a hemorrhage after a car crash. Gonzalo, the youngest son and another hemophiliac, died at the age of 19, also from an uncontrollable hemorrhage after a car accident. Jaime, the second son, who had been deaf since a childhood operation for mastoiditis, renounced his claim to the throne on the grounds of his disability. The remaining son, Juan (the father of the present king of Spain, Juan Carlos), was the only healthy son who survived to adulthood. The number of potential heirs was therefore very limited. The Nationalists, under Franco, received strong support from Italy and Germany, and by April 1939 they were victorious. From the 1940s onward, Spain was under the influence of what Franco liked to call "national Catholicism." During World War II, Spain, although in sympathy with Hitler and Germany, remained neutral. For not taking sides and for Franco's pro-Fascist policies, Spain was ostracized by the Allies after the war, and this continued to adversely affect the Spanish economy. However, by the 1950s, Spain's economy improved as it opened its markets, and it became strategically important to Britain and the United States in the Cold War. Franco permitted the United States to build air and naval bases in Spain in exchange for economic and military aid. This helped in industrial expansion and improved the economy further. In 1955, Spain was admitted to the United Nations. From 1969 to 1973 there was an unresolved power struggle between the reformists and the conservatives. Miners and other workers went on strike, Basque terrorism increased, and Franco was aging and unwell. The situation was resolved in 1975: General Franco died on 20 November, and Juan Carlos (the son of Don Juan) was crowned King on 22 November. In the end, Franco's 30-year dictatorial regime was replaced by the liberal constitutionalism he had fought so hard against.

The influence of hemophilia during World War II

Leopold, the eighth child of Queen Victoria, was born on 7 April 1853. He was the only son to be affected by hemophilia. He bruised easily and suffered many bouts of internal bleeding. He became a chronic invalid with an abnormal posture. Victoria considered him unattractive, saying, "He is tall, but holds himself worse than ever, and is a very common looking child, very pale in face, clever but an oddity—and not engaging though amusing." The Queen was so ashamed of Leopold that he was frequently left behind while the rest of the family went on holiday. Leopold used these times to read widely, and he was one of the most intelligent of Queen Victoria's children. The Queen recognized his intellectual capacities, and when he was 24 she made him one of her private secretaries with access to state papers, a privilege denied to his brother Edward Prince of Wales. In 1881, Queen Victoria made Leopold the Duke of Albany, and the following year he married Princess Helena of Waldeck, sister of the Dutch queen.

The Duke and Duchess of Albany had a daughter, Princess Alice (a carrier of hemophilia), and a son, Charles Edward Leopold, who was born shortly after his father's death at the age of 31 from a cerebral hemorrhage after falling down a flight of stairs. Charles Edward Leopold could not inherit hemophilia from his father (see p. 9–10), but at the age of 16 he did inherit the dukedom of Saxe-Coburg and Gotha from his uncle Alfred, Duke of Edinburgh, and he rose to the rank of general in the German Imperial Army. With the collapse of Germany after World War I, he had to abdicate his dukedom; however, he was a prominent member of the Deutsches Nazionalist Volks Partie (DNVP) and helped forge an alliance between it and the Nazional Sozialist (Nazi) Partie. In 1933, when Adolf Hitler was elected chancellor, the DNVP disbanded and many of its members joined the Nazis. Charles Edward Leopold became a group leader in the Nazi Brownshirts. In 1936 Hitler sent him to Britain in his capacity as President of the Anglo-German Fellowship; his real mission was to assess the possibility of an alliance between the two countries. During his stay he met with his cousin Edward VIII and several high-level politicians. After his first meeting with Edward VIII in early 1936, Charles Edward Leopold reported to Hitler that the new King would be amenable to such an alliance and would exercise his influence to make this a "guiding principle of British foreign policy." The Duke went so far as to report that Edward dismissed the need to discuss this with the Prime Minister, Stanley Baldwin. However, the Duke was not a reliable reporter, and his assertion was never confirmed by Leopold von Hoesch, the German ambassador to Britain.

In December 1936, Edward VIII abdicated (becoming thereafter the Duke of Windsor) to marry Wallis Simpson, an American divorcée. The following year, the Duke and Duchess of Windsor visited Germany, where the Duke met with Hitler, Goering, and Goebbels and greeted Hitler with a Nazi salute; his cousin Charles Edward Leopold celebrated the visit with a gala dinner in his honor. The cordial relationship between Hitler and the Duke of Windsor caused great consternation to the new King, George VI, and his government in the years before the outbreak of World War II. Early in the war, Germany unleashed a destructive air offensive against Great Britain. This was perceived by the Prime Minister, Winston Churchill, as a prelude to an invasion of Britain. The Duke of Windsor was reported to have recommended ending the war "before thousands more were killed or maimed to save a few politicians." It was even rumored that Hitler was so intent on maintaining the Duke within reach of Germany that he contemplated either kidnapping or bribing him. In order to remove Edward from any German influence and to avoid having him in Britain, Churchill arranged for him to be appointed Governor of the Bahamas. In late July 1940, Hitler, through intermediaries, audaciously suggested that the Duke might return to the British throne if Germany was victorious; there is no evidence that the Duke responded to this proposal. The Duke hesitated before accepting the Governorship of the Bahamas, partly because of German-inspired rumors that the British Secret Service was planning to assassinate him once he arrived there. He eventually accepted the post and sailed for Nassau in August 1940.

Meanwhile, Charles Edward Leopold provided encouraging, if greatly exaggerated, reports to Hitler on the pro-German party's strength in Britain, reinforcing Hitler's hope that Britain might still be persuaded to form an alliance with Germany. It has been suggested that Hitler delayed his attack on Britain in 1940 as a result. For his pro-Hitler role, Charles Edward Leopold paid a heavy price: two of his sons died in the war, and he lost most of his estates, which were in the Russian zone of Germany. After the war, the Duke and Duchess of Windsor lived in exile in France for the rest of their lives.

Consequences

Victoria and Albert were prolific parents, and their offspring married into almost all the royal families of Europe, which were notorious for marrying among themselves. Queen Victoria sowed the seeds that ultimately led to the demise of several European dynasties. Since the hemophilia gene is passed as a recessive rather than a dominant gene, and since affected individuals have tended to die young until fairly recently, this gene has been eliminated from most of Queen Victoria's descendants. The growth of constitutional monarchies and republics means that the gene no longer plays a role in world affairs. The gene may still exist in some descendants of Prince Leopold, Duke of Albany, and of Queen Victoria Eugenie of Spain, but these descendants are all commoners and will not contract dynastic marriages with other descendants of Queen Victoria. Porphyria, on the other hand, as a disease transmitted via a dominant gene, is still being passed from one royal generation to the next, although in none of these persons has it had the profound effects that it had on the Houses of Stuart and Hanover. Unless the present House of Windsor harbors an as yet undetected carrier, the royal curse of porphyria may have finally reached the end of the line.

Porphyria and hemophilia illustrate how closely intertwined are disease and culture. Appreciation of this nexus can enable a better understanding of and responses to other illnesses such as the potato blight (chapter 2), the emergence of drug-resistant tuberculosis (chapter 7), and the pandemics of influenza (chapter 10) and AIDS (chapter 11). When we consider the world of the past, a time when there were no possible interventions for the "inborn errors" of porphyria and hemophilia, we can see how a disease can dramatically affect the lives of millions of people for decades. And illness, contagious or not, when present at a critical time, can affect the conduct of wars, the nature of immigration policies, the manner in which the sick are regarded by the remainder of society, and the political fortunes of nations; it can also serve to promote new strategies to protect the public health while at the same time ensuring that individual liberty is not compromised. The stories of porphyria, hemophilia, and President Kennedy's Addison's disease are tantalizing: they tempt speculation on how events in the world during the past 150 years might have turned out differently if political leaders and prominent members of royal families had not been plagued with defective genes.

2

The Irish Potato Blight

Grosse-Ile, lying in the middle of the St. Lawrence River, is a picturesque island with a background of majestic peaks. Grosse-Ile is also a place where thousands upon thousands of men, women, and children were detained and died. The Grosse-Ile station came into being when reports to the colonial government in Canada told of sick people from the Old World, especially the Irish, who were about to arrive via the St. Lawrence River. In response, the Assembly of Lower Canada (as Quebec was then called) passed a resolution on 23 February 1832 that made Grosse-Ile a detention station for sicknesses commonly believed to originate "in the homes of the human riff-raff."

The writer Susannah Moodie, who emigrated to Canada from England in 1832, described her own impression:

> "I looked up and down the glorious river; never had I beheld so many striking objects blended into one mighty whole! Nature had lavished all her noblest features in producing that enchanted scene. The rocky isle in front, with its farmhouses at the eastern point, and its high bluff at the western extremity, crowned with the telegraph . . . the middle space occupied by tents and sheds for cholera patients and its wooded shores dotted over with motley groups added to the picturesque scene . . . Never shall I forget the extraordinary spectacle that met my sight the moment we passed the low range bushes which formed a screen in front of the river. A crowd of many hundred Irish emigrants had been landed during the present and former day and all . . . men, women and children, who were not confined to the sheds (which resembled cattle pens) . . . were employed in washing clothes or spreading them out on rocks and bushes to dry. The people appeared perfectly destitute of shame or a sense of common decency. Many were almost naked, still more partially clothed. We turned in disgust from the revolting

scene . . . Could we have shut out the profane sounds which came to us on every breeze, how deeply we should have enjoyed an hour amid the tranquil beauties of the . . . lovely spot."

Although the detention of immigrants at Grosse-Ile was presumably instituted as a means of protecting the public health, in actual fact it served as a vehicle for maintaining class distinctions and scapegoating the "wretched refuse" of Ireland.

In 1833, as the number of sick immigrants arriving in Canada dwindled, the Grosse-Ile detention station fell silent. A decade later, however, it once again became active due to the changes that were taking place in Ireland. Indeed, between 1845 and 1849 the population of Ireland would decline by over 2 million. Half of these would die of starvation, disease, and malnutrition, while the other half would emigrate. The United States was traditionally the route for Irish immigrants, but in 1847 the United States enforced an increase in the cost of passage and ships that were overloaded were to be confiscated. This opened up new routes to the United States from Canada as ship owners sought a cheaper option. Hundreds of thousands of Irish were crowded aboard unsanitary sailboats unfit for transporting human beings. During the voyages of these "pest ships," people who died, along with their possessions, were hastily wrapped in canvas and thrown overboard as if they were dead birds or garbage.

Although ships usually took 45 days to cross the Atlantic Ocean, 26 of those that set sail in 1847 took over 60 days to reach Grosse-Ile. In 1847, over 5,000 people died en route, and a like number were buried in a mass grave on the island. Four physicians at Grosse-Ile, aided by a crew of eight, worked from dawn until dark every day digging trenches and burying the dead three deep. By August, dirt had to be imported to the rocky island to bury more bodies. In spite of this, rats were coming off the ships to feed on the cadavers. All told, the number of deaths on Grosse-Ile probably exceeded 9,000, and many thousands more died elsewhere in colonial Canada during that "summer of sorrow." The epidemic disease that prompted the Irish to leave their homeland and choose immigration and almost certain death in the passage to a foreign land was the disease known as "late blight."

Politics and the Great Hunger

The earliest settlers in Ireland came from mainland Europe about 6000 BC, and about 400 BC the Celts from Britain and Europe arrived. In about

AD 400, St. Patrick introduced Christianity to this island, along with the Roman alphabet and Latin literature. Ireland, save for fish, is without any valuable natural resources, such as gold or silver, gemstones, oil or natural gas, or iron ore, and it has an unproductive soil and miserable weather. About AD 795 the Vikings began raiding Ireland. At first the people of Ireland could do little to defend themselves, since until AD 1000 the lack of iron (as well as copper and tin) left Ireland a Stone Age economy. In 1014, the Irish King Brian Boru organized the princes of several kingdoms and drove out the Vikings. Beginning in 1160, the Normans increasingly controlled Ireland, and by 1300 they were in complete control. In 1534 Henry VIII began to try to regain Ireland from the Normans, and in 1542 he forced Ireland's parliament to declare him King of Ireland. Henry tried to introduce Protestantism into Ireland, but without much success. Throughout the later 1500s Henry VIII's daughter Elizabeth I strengthened the English hold and attempted to establish Protestantism. Elizabeth I outlawed Roman Catholic services and executed a number of bishops and priests. The porphyric King James I (see p. 2), who succeeded Queen Elizabeth I, continued to seize land in Ireland and give it to the English—a system known as "plantations." This occurred especially around Ulster and established the majority of Protestants, whose descendants still live in Northern Ireland.

Under the Protestant King William III (who reigned from 1689 to 1702 and succeeded James II), Penal Laws were instituted. These laws barred Catholics from the military, commerce, and civic office; they were also denied the vote and could not purchase land. However, Catholics who converted to Protestantism were given land. In the early 1600s there were Irish revolts, but these were quickly put down by a succession of English kings; by the 1700s there was tight control by Britain. Irish Protestants objected to the restrictions, and in 1782 Britain granted autonomy to the Irish Parliament. The Catholics were given the right to hold land and to worship, but were granted no political power. In 1798, the Irish staged an unsuccessful rebellion, and although it was put down, the British Prime Minister William Pitt persuaded both the Irish and the British Parliaments to pass an Act of Union in 1801. As a result, Ireland became a part of the United Kingdom. This ended the Irish Parliament, and now Ireland would send its representatives to the British Parliament in London; later, Irish Catholics were permitted to serve in the British Parliament. During the 1800s there were several attempts to institute home rule, under which Ireland would remain a part of the United Kingdom but would have its own parliament to govern domestic affairs. The Protestants in Ulster, however, were opposed to home rule, fearing that Catholics would dominate Parliament, and so the British remained in control. In effect, Ireland became a British colony.

The attitude toward the Irish was described by one English writer: "Ireland is a little island at the edge of Europe with a Stone Age culture . . . its inhabitants wild, feckless, and charming or morose, repressed, and corrupt, but not especially civilized." Benjamin Disraeli, the beloved Prime Minister of Queen Victoria, had a stronger opinion: "The Irish hate our order, our civilization, our enterprising industry, our pure religion." Britain maintained Ireland as an agricultural colony and prevented manufacture of anything the British produced; Irish culture was suppressed save for music and dance. The British colonization of Ireland introduced a tenure system that gave Protestant landlords control of 95% of the land.

The landlords (most of whom were absentees) subdivided their land into 5-acre lots that were rented to estate agents. These lots, in turn, after being subdivided into smaller ones, could then be rented again at higher rates. At the base of this economy were 3 million tenant farmers who might have one quarter of an acre of land. On this small plot, the tenant farmer and his family cultivated a small garden and lived in a tiny one-room mud cottage with neither floor nor windows, only a door and a hole in the thatched roof to let out the smoke from the turf fire. On average, there were 10 people per cabin, which they shared with the family pig. The tenant farmers had another 5 acres that were used for growing cash crops such as wheat, oats, and barley.

The potato, introduced into Europe by the mid-18th century, was never a cash crop; however, because it was better adapted to the cool, moist conditions in Ireland than other crops, most of the Irish population was dependent on it as a supplemental food source by 1800. The Irish tenant farmers were forced to pay exorbitant taxes and to export their cash crops as well as butter, eggs, pork, and beef in order to produce sufficient income to pay the landlords for the use of the land and to avoid eviction. Some of those evicted were relocated to less fertile areas where essentially only the potato could be grown, but even these lands remained under the control of absentee British landlords. The Irish peasants on the worst land came to rely almost exclusively on potatoes to store over the winter and to feed themselves and their livestock, especially the pigs. The potato became the staff of life for the Irish peasant. A family of 10 needed a ton of potatoes per month to survive, and an average adult ate 9 to 14 pounds per day supplemented with buttermilk.

In 1845 a "queer mist" came over the Irish Sea and the potato stalks turned black as soot. The next day the potatoes were a wide waste of putrefaction, giving off an odor that could be smelled for miles. About 40% of the potato crop was destroyed. In areas where the blight was most severe, tenant farmers and their families frantically scoured the land and bogs for stray potatoes. They washed away the rotted parts and grated the remainder to make flour. Children searched the woods for nuts and berries; they dug for fern and dandelion roots and ate the leaves and bark from the trees. The streams were fished for eels and trout, and the peasants trudged many miles to get to the shore, where they scraped mussels, limpets, and seaweed from the rocks. Many died from eating poisonous plants, but "The Great Hunger" forced them to try anything that seemed edible.

When the fourth rider of the Apocalypse, Famine, rode into Ireland, the Victorian historian Charles Kingsley described what he saw: "I am daunted by the human chimpanzees I saw along that 100 miles of horrible country. I don't believe they are our fault. I believe that there are not only many more of them than of old, but that they are happier, better and more comfortably fed and lodged under our rule than they ever were." As the Great Famine gripped Ireland, thousands of Irish died each day. Few were felled from starvation alone: death invariably was visited upon the Irish because malnutrition made them more susceptible to diseases such as typhus, cholera, dysentery, and relapsing fever. Desperate to escape from virulent racism and prejudice, as well as starvation caused by the potato failure, the people of Ireland began a process of migration that changed the course of their history and that of Canada, the United States, and Great Britain.

Bite of Blight

Many of the great civilizations of the world were established by people who settled down and cultivated a staple food crop. In Southeast Asia it was rice, in Europe it was grain (wheat and rye), and for the Mayans and Aztecs in Central America and Mexico it was corn. The South American Incas cultivated a plant that grew well above the 10,000-ft level in the high valleys and cold plateaus of the Andes mountains. This plant had stems above and below ground level, and the swollen underground stems—called tubers—were highly nutritious. These tubers, which the Incas called "papas," we call potatoes. The potato tuber is nutritious because it contains proteins, starches, and vitamins. The potato was the staple food source on which the Inca civilization was built, and even today it is one of the most important food crop plants in the world. Although rice, maize (corn), and wheat are the top three food plants, the potato ranks fourth. Indeed, one-fifth of the world's people use the potato as their primary food source today.

The Incas first cultivated the potato over 6,000 years ago. When the Spanish Conquistadors under Pizarro came to the Americas in search of treasure, they destroyed the Inca civilization but discovered something more valuable than gold, silver, and jewels—the potato plant. This plant, perhaps the most priceless possession of the Incas, the Conquistadors did not even bother to record. By the late 1500s the sailors on Spanish galleons accidentally introduced these plants into Europe, where they were considered more of a curiosity than a foodstuff. They may have been rejected as a food because the potato is a member of the poisonous nightshade family. It was also claimed to cause leprosy, and even its reputation as an aphrodisiac did not make it acceptable. (The seductive Marie Antoinette, it is claimed, wore potato blossoms in her hair.) It was widely believed that eating potatoes caused flatulence, and so initially the tubers were fed to farm animals; however, since this caused no harm, the potato came to be accepted as a food fit for humans. By the 1800s, with an increase in the population of Europe and the inability of grains to support this population, the potato became a regular part of the diet. Easy cultivation and high yields in cool climates led to a major dependence on potatoes by populations on the high cold plateaus of Spain, the dank flatlands of Germany and Poland, and the soggy bogs of Ireland. The British landowners encouraged cultivation of the potato in Ireland since it saved the grain for export and for their own use. Furthermore, only in the southeastern part of Ireland is the soil suitable for growing grain (rye or wheat) from which bread can be made, but potatoes can be grown even in the poorest soils.

The Irish invented a highly efficient method of potato cultivation, called the "lazy bed." The lazy bed is one in which the seed potatoes ("eyes" of the tuber) are placed helter skelter on the ground and covered with manure and seaweed, and then the soil dug from lateral trenches is piled on top so that long, narrow beds of soil are raised 2 to 3 ft above the surrounding ground. This protects the tubers from excess moisture. The cultivation of potatoes produces a high-yield crop (~30 tons/acre).

Potato plants mature faster than most crops, taking 90 to 120 days, and edible tubers can be harvested in 60 days. The potato tuber is higher in protein than soybean, and half a potato can provide half of the human daily requirement of vitamin C.

Feudal systems tended to favor high birth rates because under such a system the number of dependents was a measure of a person's wealth. Children were a cheap and expendable source of labor and could be relied on to provide assistance in one's old age. In Ireland this led to a massive increase in the birth rate, and after the middle of the 17th century more newborns survived because there was a reduction in the death rate due to reduced infant mortality, tribal warfare, murder, and mayhem. In 1660 the population of Ireland was 500,000, but by 1688 it had doubled to 1.25 million. Between 1760 and 1840 it grew from 1.5 million to about 8 million. This explosive growth in the population of Ireland strained the economy and left many peasants living under bare subsistence conditions. As the Irish population grew, the land was further subdivided and living standards declined. The British government attempted to consolidate these small plots as a way of increasing grain output and also instituted Penal Laws which denied the Irish peasant population freedom. They were forbidden to speak their own language and practice their faith, to attend school or hold public office, to own land, or to own a horse worth more than £5. A peasant earned £7 per year, of which two-thirds was paid as rent. A pig, valued at £4 when sold, served as financial security for the peasant; it is from this that the term "piggy-bank savings" comes.

The English clergyman Thomas Malthus (1766 to 1834) wrote *An Essay on the Principle of Population* in 1798, in which he stated that a population that is unchecked increases in geometric fashion. The consequences of unrestrained population growth, in Malthus' words, would lead to "misery and vice." These would tend to act as "natural restraints" on population growth. Today we understand that there are at least two kinds of checks to set the upper limit for a population: external or environmental factors (including limited food, space, or other resources) and self-regulating factors (such as fewer births, deliberate killing of offspring, or an increased death rate due to accidents or disease). Agriculture in its most efficient form can change the environmental restraint so that far larger human populations are possible; however, even here there are limits. It is estimated that without the potato as a food source, all the land in Ireland could support a population of only 5 million if the people were fed on bread. Compounding the problem caused by potato blight, there was a worldwide shortage of bread grains at a price the Irish could afford. Between 1798, when Thomas Malthus' *Essay on Population* was published, and 1845 there were 20 failures of the Irish potato crop, all leading to starvation, disease, and debility. These were true famines. During the same period, only three famines occurred in England. According to Malthusian doctrine, any increase in the Irish population would be due to their carnal and vicious nature. Famine would control this population explosion, and in Malthusian terms this was deserved. The Irish, the British opined, were hopelessly inferior and incurably filled with vice and so they deserved the famine, which would exert control over their excessive breeding. In effect, the Malthusian theory was used to reinforce British prejudice against the Irish and to justify the British failure to provide relief. There was also a laissez faire economic policy under which the British government adopted a hands-off policy. The attitude in Great Britain was to let the market run its course. By the end of 1846, not a single potato was left in Ireland. In addition, that year had one of the coldest winters on record. The level of starvation among the people soared.

Although in 1845 to 1846 Britain's Prime Minister, Robert Peel, attempted some countermeasures, including the importation of corn for resale in Ireland, no one knew how to cook it, and in return the starving and enfeebled Irish were required to perform public work by building roads, walls, piers, and bridges. It helped only a little. Soup kitchens were started, but they dispensed essentially flavored water. The churches offered little hope since the Church of Ireland was entitled to collect taxes from tenants regardless of their religion. Indeed, the Catholic Church increased its ownership of property in Ireland during the famine. The Church was vehemently on the side of the absentee English landlords, and it was left to the Quakers to seek long-term relief for the Irish.

In late 1846 the British Parliament came under the control of a new Prime Minister, John Russell, who reduced the British financial commitment to Ireland. This placed a greater burden on landlords and private charities. The public works schemes were inefficient and bureaucratic, and the wages paid were very poor and very late. Tenant farmers held short-term leases that were payable each 6 months. If the tenants failed to pay their rent, they were evicted and their homes were burned.

There was a severe winter in 1846 to 1847. In the early months of 1847—called Black '47 (now an Irish Rock Group)—there were reports of dogs eating dead bodies in the streets. Public works projects were abandoned, and poorhouses were established. The poorhouses (also called workhouses) were mismanaged, overcrowded, and filthy, and the inmates were forced to wear prison-like uniforms. They subsisted on a watery oatmeal. To limit the number of people seeking relief, the Poor Law Extension Act was passed in 1847. This prevented tenant farmers with over a quarter of an acre of land from receiving assistance. In 1847 more than a million died of starvation or diseases such as typhus and cholera; this was a peak emigration year. Indeed, during the next 4 years, 2 million Irish emigrated from Ireland, never to return.

Conditions were made worse because the British government tried to placate the politically powerful landowners and allowed continued export of food from Ireland while preventing importation of food. The British idea of free trade led to the notion that assistance would weaken the resolve of the Irish peasants. The primary goal of the British was economic: extract the greatest amount of resources and exports from their colonies, thereby benefiting the bankers and landowners. One letter writer to the *London Times* stated that "Giving more money to Irish relief would be as ineffectual as throwing a sackful of gold into their plentiful bogs." A Chancellor of the Exchequer said, "Except through purgatory of misery and starvation I cannot see how Ireland is to emerge into a state of anything approaching quiet or prosperity." In addition, the Irish peasants were so weak from starvation and disease that they could not work the land. Compounding this problem were economic factors: the landowners, in order to meet their losses due to the famine, raised the rents of the tenants, which in turn led to nonpayment, eviction, and destruction of the houses of the tenant farmers. Between 1849 and 1854, at least 500,000 people were evicted. Thousands more were thrown out without official sanction, and homelessness became as much a problem as hunger.

At first the potato failure was believed to be due to God's anger over the excesses of the people. Later, it was shown that the failure was due to "late blight," a disease causing large necrotic areas (called blight) on potato leaves that occurs in the late part of the growing season, e.g., August and September. Late blight reappeared again in 1848 and 1849, and in some places 1849 was as bad as 1847. Many people saw emigration as their only solution. According to the Poor Laws, the landlords were to support the peasants who were sent to the workhouse. This cost £12 a year per person. Some landlords, however, economized and paid for the passage of the peasants to Canada, which cost only £6 a head! The very poor migrated to England—1.5 million went to Liverpool, London, Manchester, and Birmingham—whereas those who were slightly better off and could afford the cost of passage emigrated to the United States. Only about one-fifth of the migrants survived the trip across the Atlantic because of their poor health, the fact that it took weeks to months to cross, and no food was provided on board ship. These were not passenger ships: they were ships ordinarily used for hauling timber and cattle. There was no place to cook and no place to put the sick, and there were no proper latrines. The filth and stench below deck were overwhelming. Many of the passengers carried lice and were infected with typhus. Because of the high death rate on board, they were called "coffin ships." And it is a bitter irony that in Ireland during this period, while people were starving, grain was still being exported. The potato famine changed the structure of landholding in Ireland—the poorest were evicted, but the landlords were also financially ruined, crushed by the burden of falling income and higher taxation. Many landlords sold out to larger landowners, who in turn were also unpopular with their tenants.

The Great Hunger's Cause

The perils of a single-crop economy have seldom been better illustrated than in Ireland in 1845 to 1849, for at that time without the potato the Irish economy could not survive for very long. While other regions of Europe may have been able to turn to alternative food sources, this was not possible for the Irish. The potato blight was an ecological disaster compounded by the failure of government. Some consider it to be equal to the holocaust. Although theories as to the cause of late blight were many, including an act of God, introduction of the steam locomotive, and excessive uptake of soil water that the potato could not expel, it was the Reverend Miles J. Berkeley who in 1846, after making careful microscopic examination of diseased plants and seeing a whitish felt on the leaf surface (resembling that in moldy bread), proposed that it was none of these but instead was a fungus. Berkeley was mocked, and his contention gained little support. Indeed, the prevailing opinion of the time was that a cold and damp miasma resulted in blight.

A critical question regarding blight was: which came first, decay followed by fungus or fungus and then decay? In 1861 the great German biologist Anton de Bary clearly showed that late blight was caused by the fungus he named *Phytophthora infestans*, the "plant destroyer." To confirm the role of the fungus, de Bary did a simple experiment: he grew healthy potato plants in pots, divided them into two groups, and deliberately dusted spores from the plants with blight onto the moistened leaves of a group of healthy plants; he left the other group ("controls") alone, making certain that spores could not reach them. Both groups were exposed to a cool, moist environment where the miasma could do its work. In a few days the telltale sign of blight—spots of decay—appeared on the leaves of the fungus-inoculated plants. The control group showed no sign of disease. Clearly, potato plants did not rot because of a miasma or because they took up too much water. de Bary suggested that the microscopic spores ride on the stormy winds and that blight results when rain splashes them on to the leaves. In this way the infection spreads from plant to plant, field to field, and country to country. The significance of de Bary's work led to a novel understanding of sickness: parasites can be the cause of a disease. Today de Bary and Berkeley are rarely recognized for their pioneering work, yet their experiments anticipated Louis Pasteur's germ theory of disease by nearly a quarter of a century.

In late blight, the signs of impending disaster first appear on the leaves in the form of brown-black spots. Under moist conditions the spots enlarge quickly and the plant has a pungent odor. A white fuzzy growth, barely visible, appears in the spots; under the microscope, these are seen to contain the long tube-like threads first seen by Berkeley. The threads (called hyphae) divide and twist like snakes to form an extensive network; they penetrate the plant tissues and act like "soda straws," allowing the fungus to "drink" the rich nourishing sap of the potato until the leaf and stem are literally sucked dry. This takes only 3 to 5 days. At the tip of each filament a swelling develops; within the swelling, microscopic spores are produced. Millions of spores can be produced on an infected leaf, and each spore is so tiny that 500 of them would be no larger than the period at the end of this sentence. The spores germinate, giving rise to either tube-like threads or swimming spores that can also form threads. The tubes enter the leaf through its microscopic pores (stomata), or the leaf tissue is eroded by digestive enzymes released by the hyphae. The cycle of spore formation and germination is favored by moist, cool conditions. But how does blight on the leaf produce rotten potatoes? de Bary buried healthy potatoes in the soil, shook the spores from blighted leaves on the soil surface, and gently watered them as if it were raining. The spores washed down, and when the potatoes were dug up, they too were blighted. Clearly at harvest time millions of spores are washed from the leaves so that the fleshy tuber itself becomes infected. Its skin is discolored with brown-purple blotches resembling bruises, and as the microscopic threads of the fungus penetrate deeper, the tuber begins to rot. Dry rot of tubers by *Phytophthora* is followed by wet rot of the potato due to other microbes in the soil.

One of the mysteries concerning the late-blight fungus was how it was able to survive the cold of winter. Did it overwinter in the soil or in the tuber? Although observations of the hyphae under freezing conditions showed that they were too fragile to withstand low temperatures, within the tuber itself the fungus was protected and able to survive very low temperatures. Since only the tuber was kept through the winter, the blighted potatoes provided the source of infection for the next season's crop. Because in a single growing season it is possible to get many cycles of late blight, it is considered to be a compound-interest disease with great powers of amplification that can lead to an explosive outbreak. As a result, an entire potato crop can be quickly destroyed. Indeed, one infected tuber per 2.5 acres can cause an epidemic of late blight, especially when the weather is right: cool, with high rainfall and humidity.

Where did the "plant destroyer" come from in the first place? Most likely *P. infestans* was introduced with tubers brought to Belgium from Peru. The disease was first reported in Belgium in July 1845, and it soon spread throughout Europe. Once introduced, it became a menace. Although today topical fungicides such as Bordeaux mixture (made from copper sulfate and lime and developed in the 1890s) may be sprayed on leaves before the disease begins or Ridomil, a systemic fungicide developed in the 1970s, may be applied to the soil, none of the fungicides eradicate blight—they simply reduce the amount of defoliation so that tubers can be harvested in respectable quantities. However, in 1845 to 1849, none of these were available.

Consequences

The social and political impacts of the Great Hunger or the Irish Potato Famine (1845 to 1849) were profound. The mass immigration to the United States was so great that today 1 of every 10 Americans is of Irish descent. By 1914 there were more than 5 million Irish-Americans. These immigrants were no longer agricultural peasants working the lazy beds of the bogs but were urban dwellers who came to occupy key positions in railroading, mining, civil engineering, law enforcement, and politics.

Before 1840 the Caucasian population of the United States consisted of Protestants, most of whom came from Britain, The Netherlands, Germany, and Scandinavia; a smaller number of Catholics from France, Spain, Switzerland, and Germany; and very few Irish Catholics. Although the total number of Irish Catholics who emigrated to the United States between 1810 and 1840 was less than 100,000, over the next 25 years the blight-caused famine created a flood of Irish Catholic immigrants, whose numbers exceeded 100,000 per year. These immigrants favored the cities, especially Boston and New York—the ports where they first disembarked. In these urban centers they could constitute as much as 30% of the population. These poor, unskilled laborers with few relatives or homes crowded themselves together in Irish Catholic ghettos, where the priest was their friend and counselor on Sunday and the local politician became their advisor and caregiver for the remainder of the week. The ghetto Irish constituted a voting bloc that could exercise great political influence in the cities in which they lived. Through their social activism and unionization, these immigrants, the poorest of the poor and despised by most in the United States, changed political party platforms, especially that of the Democratic Party—a legacy that persists today. Anti-imperialist policies in America (and isolationism) were fostered by the Irish immigrants largely because of their hatred of Britain. It is thought that it was the effective lobbying efforts by the Irish (which continued to denigrate British imperialism) that delayed U.S. entry into World Wars I and II. The virulent anti-English position of the Irish, the first politically and ethnically integrated group, affected Northern attitudes toward the American South. Indeed, during the Civil War and thereafter, the Irish voted for the hard-line Republicans, not because they were anti-slavery but because they hated the English and the white Anglo-Saxon Protestants (WASPs) who lived in the American South.

Could there be a repeat of the 1845 to 1849 disaster? Possibly. Potato famines, smaller in scale than that of the Great Hunger but certainly not inconsequential, have been recorded. The last major famine due to late blight occurred in 1916 to 1917 and resulted in the death of 700,000 German civilians: the old, the weak, the children. During World War I—the Great War (1914 to 1918)—the copper used in the preparation of Bordeaux mixture was needed for shell casings and electric wire, and without this fungicide, potatoes rotted in the fields. All the grain and the potatoes that remained were commandeered for the war effort, and so the civilians behind the lines were left to feed on turnips or any other foods they could find. In autumn of 1916 the people were hungry, and by winter they were starving. Germany's High Command assumed that the eastern campaign against a weakened Russian army would be simply a mop-up affair, and, once complete, the plan was to direct the army's efforts against the Allies on the Western front. That did not happen because there was a saboteur, *P. infestans*. Many of the rank-and-file German soldiers, knowing of the plight of their hungry, starving, and dying families at home, lost their will to fight. It has been speculated that this weakening of morale was one of the reasons the German High Command was never able to launch a successful campaign on the Western front. Indeed, the German army was forced to retreat and maintain their position on the Hindenberg line. In 1917, British and French forces were joined by the Americans, and a great new military strength could be brought to bear on the German front. But that strength became unnecessary since late blight had already struck a devastating blow against the German Empire. With morale in decline, the military might of Germany began to crumble, and by 1918 it collapsed. The Great War ended when the armistice was declared on 11 November 1918.

Even today in countries such as Russia, where for many people there is little to eat except potatoes, an epidemic of late blight could be catastrophic. When the blight appeared in the 1990s, yields from some Russian plots were reduced by as much as 70%. In countries where fungicide is affordable and is applied, losses can still be as high as 15%. It has been estimated that in developing countries, where fungicides are out of reach because of cost and difficulties in distribution, the annual toll amounts to billions of dollars. What to do? At a minimum, it is necessary to continue to search for new potato varieties that are resistant to blight and develop more effective fungicides, but more importantly it must be recognized that preventing future famines will require both political will and global social responsibility.

3

Cholera

Cholera is a horrible disease. At first, the symptoms produce no more than a surprised look as the bowels empty without any warning. Then surprise changes to agony as severe cramping pains begin. Copious quantities of liquid, resembling rice water, pour through the anus. As the pain intensifies, the only small relief is to draw oneself into a ball, chin held against the knees; the breath whistles softly between the teeth. When death occurs at this stage, the body cannot be unrolled and the victim has to be buried in the fetal position. Those who do not die from this first attack suffer a slow and painful decline. The cheeks become hollow, the body liquids surge more slowly but still remain beyond control, and the watery stools contain fragments of the intestinal lining. As the hours pass, the skin darkens, the eyes stare vacantly without comprehension, and then life ends.

The historian William McNeill described it this way:

> "The speed with which cholera killed was profoundly alarming, since perfectly healthy people could never feel safe from sudden death when the infection was anywhere near. In addition, the symptoms were particularly horrible: radical dehydration meant that a victim shrank into a wizened caricature of his former self within a few hours while ruptured capillaries discolored the skin, turning it black and blue. The effect was to make mortality uniquely visible: patterns of bodily decay were exacerbated, as in a time-lapse motion picture, to remind all who saw it of death's ugly horror and utter invincibility."

For more than two centuries, cholera has outflanked the best public health defenses, and even today it remains a threat to many of the world's people. The most recent cholera pandemic began in the Celebes, Indonesia, and has relentlessly moved across the globe. It was in the Middle East in the 1960s and spread across Africa in the 1970s. In 1991, after an absence of almost 100 years, cholera reappeared in the Americas. The first confirmed cases occurred in Peru, and by year's end there were nearly 400,000 cases and 4,000 deaths worldwide. In 1992 there was a mild outbreak involving 76 people aboard a flight from South America to the United States due to undercooked or raw fish and vegetables. In 1994 there was a cholera outbreak in Goma, Zaire, that killed 50,000 of the half-million Rwandan refugees, mainly Hutus, who were escaping from the Tutsi rebels. This outbreak lasted only 21 days! And after nearly a decade without a cholera epidemic, Angola, an oil-rich nation in Africa, suffered its worst outbreak; in the first 5 months of 2006, 43,000 had sickened and 1,600 had died.

Cause

Prior to 1850, cholera was believed to travel in the air as a miasma, breathed in by all. Max von Pettenkofer, head of the Hygienic Institute in Munich, claimed that the foul-smelling and poisonous miasma rose from the low-lying marshes and polluted streams to cause disease. When cholera broke out in 1854, he sought to identify the critical environmental factors. He made a map of where the victims lived (and died) and found that neighborhoods in low-lying marshlands had the highest incidence of disease. Clearly, he reasoned, the disease must be associated with the soil. He proposed that clean air entered the damp soil, where it initiated a chemical reaction that resulted in a poisonous vapor, and people who lived nearby came down with cholera. However, John Snow (1813 to 1858), a prominent London physician and personal physician to Queen Victoria, drew an entirely different conclusion. Snow had attended many patients with cholera, yet he never came down with the disease. He asked himself how this could be so if cholera was airborne. He observed that cholera always seemed to affect the gut before the patient fell generally ill; this suggested to him that cholera probably resulted from ingestion of a poison that reproduced itself in the gut and was not due to an airborne miasma. The poison, he surmised, was released with the feces, and this in turn contaminated the water supply.

In 1849, Snow summarized his ideas in a small pamphlet, *On the Mode of Communication of Cholera*. Included in the second edition of the pamphlet, published in 1855, was a discussion of the 1854 cholera outbreak in London's Golden Square neighborhood. Snow (like Pettenkofer) made a map of the affected houses and recorded the number of deaths and the number of survivors in each building. Despite the similarity in mapping techniques, Snow's conclusions on the cause of the Golden Square epidemic were in distinct contrast to those of Pettenkofer. Snow wrote, "I found that nearly all the deaths had taken place within a short distance of the [Broad Street] pump. There were only ten deaths in houses situated decidedly nearer to another street pump. In five of these cases the families of the deceased persons informed me that they always sent to the pump in Broad Street, as they preferred the water to that of the pump which was nearer." Snow took a sample of water from the Broad Street pump and, on examining it under the microscope, found that it contained "white flocculent particles." He was convinced that these "particles" were the source of the cholera infection. On his suggestion, the handle of the pump was removed. The spread of cholera stopped. Investigation of the Broad Street pump revealed that the well below the pump was 28 feet deep and was located within yards of a sewer. Snow suggested that the well had been contaminated by sewage either from a broken sewer line or from nearby cesspools and that over time the pump drew more and more sewage; the result was a cholera outbreak. His postulate that contaminated water (drawn by the Broad Street pump) was the source of cholera is often used to illustrate the strength of statistics (and epidemiology) in identifying the manner by which a disease moves through the community; however, by the time the pump handle was removed, the cholera epidemic in Golden Square was already in rapid decline. In spite of this, Snow's general conclusions remain sound, and in time his body of work would provide the basis for preventive measures.

In 1849 there was a cholera outbreak in London, and in this case Snow traced the incidence of the disease to two water companies that had supplied the city with water. The companies drew their water from the same source, the River Thames, and both companies had their intake pipes located downstream of a sewer outflow pipe. Snow had clearly pointed to the fact that cholera was a contagious disease carried in the water, and when a subsequent outbreak occurred in 1854, Snow had an opportunity to conduct what he called "The Grand Experiment." Of the two companies that had supplied water to London, one, the Lambeth Water Company, had moved its intake upstream of the site where sewage was deposited in the Thames. Snow found that most of the affected individuals drank water supplied by the Southwark and Vauxhall Water Company whereas fewer cases were found among people who drank water supplied by the Lambeth Water Company. Since the former company drew its water from the Thames close to the outlet of the sewage system—it was malodorous and contained organic materials—it was suspected that this was the source of the cholera. Although John Snow never discovered the "germ" of cholera, he was able to correctly conclude, through meticulous gathering of data, statistical analysis, and inductive logic, that contaminated water was the source of this disease. He recommended frequent washing when around an infected person, cleaning of soiled bedding and clothing, boiling of water, isolation of the sick, and maintenance of the water supply free of sewage and cesspool contaminants. Despite his recommendations, cholera outbreaks continued, largely because most physicians were not convinced of the cause and no one had actually seen the "germ" of cholera.

In 1857 Louis Pasteur, the French chemist who demonstrated that fermentation (in the production of beer and wine) required a living microbe ("germ") and that milk and wine soured because of the presence of microbes, stated that when "germs" invade the body, they can cause human diseases. Pettenkofer dismissed Pasteur's germ theory; cholera, he said, had nothing to do with spoilage, fermentation, and microbes but instead was caused by a miasma. Then in 1882, a fellow German, Robert Koch, caused a stir when he claimed that the soils of Bombay and Genoa where cholera occurred did not meet Pettenkofer's specifications. In 1883, when cholera broke out in Egypt and thousands were dying in Cairo, Koch (not Pettenkofer) was dispatched to Egypt. He had the intuitive insight to suggest that the disease was caused by a bacterium which produced a special poison and that this toxin caused the profuse, watery diarrhea. Obtaining fecal samples and examining these microscopically, he discovered a comma-shaped bacillus which he called *Vibrio cholerae* because of its vibrating wiggles. The French had also sent a medical team from Pasteur's laboratory in Paris; they arrived in Egypt at about the same time as Koch, took samples of blood from cholera victims, and injected the blood into rats and guinea pigs. No disease resulted. When one of the members of the team came down with cholera and died, it so unnerved his colleagues that the research was abandoned and they returned to Paris. German science was victorious over French science.

Koch went a step further. He defined a strategy (called Koch's postulates) for unambiguously identifying the causative agent for a microbially induced disease. First, it must be shown that the agent is present in every case of the disease. Second, the agent must not be present in other diseases. Third, after isolation and repeated growth in pure culture, the agent must produce the same disease when introduced into a healthy animal. Fourth, the agent must be reisolated from the experimentally infected animal. If these conditions are satisfied, then a specific microbe must cause that disease.

Koch was celebrated as a hero when he returned to Berlin on 2 May 1884. Max von Pettenkofer, the ardent proponent of the miasmatic origin of cholera, insisted that *V. cholerae* did not cause the disease, and he asked Koch to send him a culture prepared from a patient who had died of the disease. Koch sent him a flask of culture, and Pettenkofer, after estimating the number of bacteria to be a billion, put the flask to his lips and drank it in front of an audience made up of his adoring students. Pettenkofer did not become ill. He wrote Koch: "Herr Doktor Pettenkofer presents his compliments to Herr Doktor Professor Koch and thanks him for the so-called cholera vibrios which he was kind enough to send. Herr Pettenkofer has now drunk the entire contents and is happy to inform Herr Doktor Professor Koch that he remains in good health." Koch did not reply. (Pettenkofer failed to mention that he did have diarrhea and that he had suffered a bout of cholera some years earlier. Perhaps the reason he did not die was his partial immunity by previous exposure to the disease.) This one demonstration, heroic and dramatic as it was, was insufficient to refute the germ theory of disease, i.e., that microbes which invade the body cause disease.

Containment

In the 1700s and 1800s the water in urban centers in Europe and America was so filthy that only the poor drank it, and they died of cholera. How did the water become poisonous? When human populations were much smaller and people lived in villages, water was drawn from the nearby streams and rivers or shallow wells were dug. Feces—human excrement—were disposed of in a privy or by being spread over fields, transforming it into "night soil." The system worked reasonably well until urbanization and industrialization came about in the 18th century. With the Industrial Revolution, the human population soared, villages became towns, and towns became cities; as there were fewer places on the land for the "night soil," its disposal became more difficult. The solution was to dump it into the rivers and streams, where it would be carried away by the current. The swifter the flow of water, the quicker the pollutants were removed, but sometimes the rivers were tidal and flow was intermittent. Also, the same rivers and streams that served for waste removal were also the source of water for drinking, washing, and bathing. The textile and steel mills of the expanding 18th and 19th century Industrial Revolution required more and more factory workers, creating an in-migration from the villages to the city. To accommodate these burgeoning numbers, row houses were built; these urban dwellings were arranged back-to-back, separated by the narrowest of streets and alleys, thereby using the smallest amount of land for the largest number of people. Families, consisting at times of 10 or more people, occupied a single room, and the row houses had a common outside water supply and privy. Under such crowded conditions there was little privacy, cleanliness was difficult to achieve, and clean water was often unavailable. Privies were used by dozens of people and were rarely cleaned out, and wastes overflowed into the surrounding back alleys and the shallow wells. Thus, "every evil of the industrial town originated in the agricultural village and resulted from the transfer not of people only but of their manner of living. This way of life did not become intolerably dangerous in the countryside because communities were small, cottages widely separated, and work performed in the open air. The disaster of epidemic illness became inevitable when the community was cramped for space, houses lay cheek by jowl and factory hands worked close together for long hours in an enclosed atmosphere." In this way, the urban dwellers in Europe and the Americas came to be deprived of the benefits of hygiene.

For over a century the miasma of cholera was believed to be God's punishment of those who sinned by drinking, overeating, or indulging in sexual excess. Sinners succumbed, the clergy claimed, because the vital forces had been weakened and dissipated. Not only was the individual at fault, but cholera also signaled that the entire nation was in moral decline. This interpretation of cholera did not "imply the impossibility, the impiety, or even the undesirability of attempting to explain and prevent disease. Indeed, both theologians and physicians agreed that it was man's duty to employ God's temporal means to preserve human life; prayer could not be expected to prevail if man did not alter the causes through which cholera acted." The belief was that through social organization and scientific inquiry, cholera could be vanquished. It is ironic that although the dominant theory for the causation of cholera (a miasma) was wrong, physicians and social reformers did institute measures to improve the public health. The miasmatist Max von Pettenkofer was of the opinion that the solution to cholera was not cure but prevention. Cleaning up the water, as well as providing good food and fresh air, would restore health. Appealing to German national pride, he called on his countrymen to demand clean water. He proposed the building of an aqueduct that would carry freshwater from the foothills of the Alps to the people of Munich. By 1865 the great aqueduct was finished. Pettenkofer urged that there be adequate supplies of water not only for drinking purposes but also to keep the streets clean so that the miasma-breeding filth would not accumulate. This too was accomplished, and soon the people of Munich were enjoying better health; however, it had nothing to do with Pettenkofer's efforts to control the miasma.

Edwin Chadwick (1800 to 1890), a lawyer-journalist, was Pettenkofer's equivalent in Great Britain, and he led the movement to improve sanitary conditions in his own country. Chadwick was appointed Secretary of the Poor Law Commission in 1832, and over the next 20 years he became the most powerful member of the General Board of Health, which had sweeping powers: house-to-house visits in search of infectious diseases, removal of the sick from overcrowded tenements, vaccination of smallpox contacts, investigations into the causes of disease outbreaks in schools and factories, inspection of sanitary improvements, and determination of the causes of unexpected deaths.

In 1842 he produced a document entitled *Report into the Sanitary Conditions of the Labouring Population in Britain*, demonstrating that the life expectancy in the cities was much lower than in the countryside. Chadwick and his Boards of Health produced "sanitary maps" showing the relationship of disease—especially cholera—to overcrowding, lack of drainage, and defective water supply. Chadwick was not entirely altruistic: he believed that a healthier population would be able to work harder and would cost less to support. He was not a physician, and so he turned to physicians and sanitary engineers to assist him. In the 1840s, piped water was already in use in many cities, but the companies that supplied the water used bored elm trunks as conduits; since these split under the pressure required to move water efficiently into the city, the water had to be supplied at very low pressure by ground-level standpipes. Later it was shown that iron pipes could support improved pressure, and water flow was increased. Next came the problem of sewage disposal. The old fashioned "sewer of deposit," i.e., cesspools and privies, was replaced by the use of narrow-bore self-cleaning drains (instead of the brick channels with stone covers), but these drains required a constant flow of water to be cleansed and to prevent blockage. This led to the development and installation of flush toilets (patented in 1819) instead of privies. The flush toilet did provide a more sanitary means of disposing of wastes: however, initially it was a public health disaster since it was not recognized that, almost without exception, the source used for drinking water was the same river used for the outflow from the toilets.

Chadwick came to be hated for his Board of Health policies and the Poor Law Guardians, which had the sweeping powers of "cleaning stagnant pools and ditches, inspecting lodging houses and prosecuting any person failing to abate nuisances," as well as his demand for vital statistics on deaths and numbers of cases of infectious diseases from the Unions. In 1854, unable to arrest the cholera epidemic in London, Chadwick and the other members of the Board of Health were dismissed. Despite his and the Board's unpopularity, some of the measures he instituted—paving roads, cleaning the streets, providing clean water, and carrying off wastes—contributed to improving the public health. And, like so many of his time, Chadwick was under the impression that cholera was caused by air pollution!

Quarantine

Isolation of contagious individuals—quarantine—has been used since the time of the Greek physician Hippocrates (ca. 460 to 370 BC). In the 19th century, quarantine involved the inspection of ships, cargo, and passengers for evidence of contagious diseases (see p. 73). Inspections were conducted at an offshore quarantine station where ships were berthed and the passengers were examined for cholera, typhus, smallpox, leprosy, yellow fever, and plague. Those found to harbor these diseases were admitted to an isolation hospital at the quarantine station, where they were attended by hospital staff. Since there were no anti-infective drugs or antibiotics, treatment was supportive: bed rest, fluids, food, and pain-relieving medications. Although quarantine has been and still can be effective as a control measure (witness the SARS epidemic of 2003), more often than not it has been put in place primarily to reassure the public that steps are being taken to prevent the spread of disease. When quarantine becomes a social policy, its effects can be pernicious: it can isolate more than those labeled as "diseased" and can stigmatize an entire group.

In June 1892, cholera came to famine-stricken Russia; it is estimated that there were 600,000 cases and 300,000 deaths during that year. Two months later the epidemic reached Germany, and by August the city of

Hamburg was dealing with a cholera outbreak. The port of Hamburg was, at the time, the largest in the world, and each day dozens of ships would depart for other parts of the world. One of those destinations was New York City, the port of disembarkation for 75% of immigrants coming to the United States. Nearly 10% of all Eastern European immigrants were Jews, who were particularly anxious to emigrate because of the harsh living conditions and persecution in Czarist Russia. When news of a mass exodus of Jews from Russia reached America, it was feared that these immigrants would sow the seeds of a cholera epidemic.

The 1892 quarantine began with a jurisdictional confrontation between the two agencies that existed to protect the public health: the federal agency (Marine Hospital Services) and the New York City Health Department. The federal quarantine was to be for 20 days but would apply only to steerage immigrant passengers, and not cabin passengers, even if they originated from the same cholera-ridden port. This quarantine had nothing to do with the scientific understanding of the cause or mode of transmission of cholera but instead was an attempt to halt steerage-class immigration, since a steamship could ill afford to sit idle in quarantine at a cost of $5,000 per day due to a loss in income and fees incurred in caring for the detained passengers as required by law. As a consequence, steamship companies would be forced to discontinue carrying steerage passengers. The federal policy (described in an executive order) was promulgated by President Benjamin Harrison, who characterized immigrants as a "direct menace to the public health" and who sought to "restrict the immigration of Russian Hebrews" into the United States. Under the federal quarantine, the epidemic would be managed by politics and nativist sentiments rather than good public health practices. The health officer of the Port of New York balked at this federal intrusion into states' rights and proposed that a much shorter 5- to 8-day quarantine period would be sufficient. Eventually a compromise was reached: ships leaving port on or before 1 September fell under the jurisdiction of the Port of New York, and those leaving after that date would be under the federal Marine Hospital Services quarantine.

Five ships (that had departed Hamburg before 1 September) arrived in New York between 31 August and 15 September. One of the first ships to arrive was the SS *Moravia*, on which there had been 22 deaths prior to arrival in New York harbor. These passengers acquired cholera during the voyage by drinking water that had been drawn from the contaminated River Elbe at Hamburg. The ship was quarantined until 23 September, when three additional deaths were recorded. A second ship, the SS *Normannia,* arrived on 3 September with a crew of 300, 482 immigrants in steerage, and 573 cabin passengers. Of the more than 1,300 passengers and crew placed in quarantine, with 36 suspected cases, 13 died of cholera (5 had already died at sea). The third ship, the SS *Rugia,* arrived on 3 September, and among the passengers in quarantine there were 42 cholera cases; 12 of these infected individuals died; 4 persons were reported to have died at sea. The ship was released from quarantine on 8 October. The fourth steamer, the SS *Scandia,* arrived on 9 September, with 26 passengers suspected of having cholera; 13 died in quarantine (32 had already perished at sea). The ship was released from quarantine 20 days after its arrival in New York harbor. The last steamer, the SS *Bohemia,* arrived on 15 September; it was carrying several hundred Jewish immigrants in steerage, among whom there were 19 active cholera cases (there had been 11 deaths at sea), and under quarantine there were 10 deaths. All told, 60 immigrants, mostly East European Jews, died of cholera at the quarantine station and 76 died en route.

In 1892 most Americans considered cholera to be a foreign disease. The public clamored for something to be done both at the quarantine station and on shore. By 4 September, inspections of more than 39,000 tenements on the Lower East Side, an area where many Jewish immigrants lived, had begun. Physicians, sanitary inspectors, and police would arrive at a home where there was suspicion of a diarrheal disease. Once a case was reported, a black-wagoned horse-driven ambulance arrived at the scene and rubber-suited sanitarians entered the home and examined those suspected of harboring disease. Those who died were wrapped in sheets saturated with bichloride of mercury disinfectant and removed for autopsy. The homes suspected of being rife with cholera or some other diarrheal disease were disinfected, quarantined, and blockaded by the sanitary police. The names of the afflicted and their families were published in the daily newspapers. And what did all this produce aside from panic and rumor mongering? Eleven cholera cases and nine deaths! The home inspections turned out to be time-consuming blind alleys that provided no clear pattern of transmission or distribution or any information about how the cholera cases came into New York City.

In 1892 a prominent New York physician said, "The history of every cholera epidemic in this country has proven that the disease entered our port on account of defective quarantine and it has been carried to us mainly by filthy immigrants." The rapid isolation of cholera victims by the

New York City Health Department and the quarantine of suspected cases may have been a factor in limiting the spread of cholera within the city. In addition, the Croton aqueduct water filtration and supply system, which provided fresh, clean, and cholera-free water in New York City, may have protected the public health of the city's inhabitants during the crisis. Although the final total of cholera cases and deaths during the 1892 epidemic in New York was smaller than that during previous epidemics in that city or in Europe, the public health authorities declared the quarantine to be a success aided by the value of "bacteriological examination." The death rates in quarantine and the emergence of cholera within the city proper give reason to question this. The management of the epidemic depended more on class distinctions and national origins than on bacteriological principles. Although immigrants were disinfected and ill persons were removed to quarantine stations, inspections were often cursory, sanitation at the quarantine station was poor, and the ill mingled with the healthy. Indeed, many obstacles to an effective quarantine existed: there were administrative problems created by a separation of public health powers among the federal and local authorities; there was a lack of detention facilities and hospitals; there was overcrowding; the facilities for laundry, dining, and sleeping were poor; the water supply at the quarantine station was unprotected and not monitored for contamination; the privies were filthy and contaminated with feces; and the sanitary habits of the immigrants themselves left much to be desired.

The 1892 cholera epidemic had other, more pervasive consequences. The cholera epidemic in Europe and Russia was the nucleating factor for American misgivings about unrestricted immigration into the United States, and the stigma of disease was cast over all European Jews seeking entry into the United States long after the threat of cholera had ceased. The quarantine also had an immediate effect in curbing immigration: in the first 9 months of 1892, the average number of Jewish immigrants entering the United States was 3,000, but for the remainder of the year it averaged 250.

The Legend and Legacy of Florence Nightingale

The legend of Florence Nightingale—the "gentle ministering angel" and "sanitary crusader in the cause of hygiene"—has on recent occasions received stinging attacks, yet a reexamination of her contributions during the Crimean War and thereafter shows that her lifetime work was instrumental in improving the public health, providing for better design of hospitals and management of patient care, and pioneering the use of medical statistics.

Florence Nightingale (1820 to 1910), the Lady with the Lamp and the founder of modern nursing as we know it today, began her nursing career at the Institution for the Care of Sick Gentlewomen in London in 1853. She was an early disciple of Edwin Chadwick, and although she had no concept of bacteria or viruses, she clearly understood the nature of contagious diseases, believing that illness in her patients was the result of "filth in which they lay, the air they breathed, the water they drank and the food they ate." Experiences as a volunteer nurse during the cholera outbreak in London in 1854 convinced Nightingale that the then current medical treatments—"infusions of arsenic, mercury, opiates and bleeding"—hastened death rather than saving lives. This background and philosophy would stand her in good stead when she was appointed Superintendent of Nursing for the Army Military Hospitals during the Crimean War (1854 to 1856).

Within a week of her appointment, she had recruited 38 nurses. When this small nursing contingent arrived at the British Army Base in Scutari (on the outskirts of Istanbul, Turkey), the two hospitals were already caring for 3,000 soldiers wounded or suffering from disease. Nightingale wrote,

> "The men after receiving summary treatment as could be given them at smaller hospitals in the Crimea . . . were forthwith shipped in batches of two hundred across the Black Sea to Scutari. This voyage was in normal times 4 days and a half; but the times were no longer normal and now the transit lasted for a fortnight or three weeks. Between and sometimes on the decks the wounded, the sick, the dying were crowded—men who had undergone amputations, men in the clutches of fever or frostbite, men in the last stages of cholera . . . without bedding, sometimes without blankets and often hardly clothed . . . There was no food beside the ordinary rations of the ship and even the water was sometimes so stored that it was out of the reach of the weak."

For many months the average mortality rate during these voyages, called the Middle Passage, was 74 in 1,000. Overall, the mortality rates during the Crimean War were horrific: one of every five men sent to war died there, mostly from infections.

The sanitary conditions during the ship passage from the Crimea were abominable, and those at the Scutari hospitals were not much better: "The wounded lying on beds placed on the pavement itself were bereft of all comforts; there was a scarcity alike of food and medical aid; fever and cholera were rampant, and even those comparatively slightly wounded, and should have recovered with proper treatment, were dying from sheer exhaustion brought by lack of nourishment they required." In short, the hospital was a disaster. "Huge sewers underlay it and cesspools loaded with filth wafted their poisons into the upper floors. The stench was indescribable. The wooden floors were so rotted that it prevented them from being scrubbed; the walls were thick with dirt; incredible amounts of vermin (rats, fleas and lice) swarmed everywhere. The building contained 4 miles of beds spaced so close that there was barely enough room to pass between, and there was no ventilation. The sheets were of canvas and so coarse that the wounded men recoiled from them begging to be left in their blankets; there were no basins, no towels, no soap, no brooms, no mops, no trays, no plates, no slippers nor scissors. There were no knives or forks or spoons and heating fuel was in short supply. Stretchers, splints and bandages and ordinary drugs were lacking."

Within 10 days of her arrival, Florence Nightingale reorganized the kitchen in the hospitals. The ill-cooked hunks of meat, served at irregular intervals, which had been the only diet for the sick, were replaced by punctual meals, well prepared and appetizing, while extra foods—soups, wines, and jellies—were also provided. When Nightingale arrived, diseases such as cholera and typhus were so rife in the hospital that injured soldiers were seven times more likely to die of these diseases than on the battlefield. Nightingale's facility with mathematics allowed her to calculate that an improvement in sanitary conditions would result in a significant decrease in the number of deaths; by February 1855 the mortality rate had dropped from 60% to 42.7%. Nightingale used statistical data to graphically represent mortality. For example, during January 1855, deaths in British field hospitals included 2,761 deaths from contagious diseases, 83 from wounds, and 324 from other causes; the army's average manpower for that month was 32,393. Using this information, she computed a mortality rate of 1,174 per 10,000 and suggested that if this rate continued and troops had not been replaced frequently, disease alone would have killed the entire British army in the Crimea. By the spring, as a result of the cleaning out of sewers, establishment of a freshwater supply, and introduction of fruits and vegetables into the diet, the mortality had dropped to 2.2%.

At the time (20 years before Pasteur and Koch promulgated the germ theory), Nightingale's sanitary interventions were revolutionary: "She and her nurses washed and bathed the soldiers, laundered their linen, gave them clean beds to lie in, fed them, while working and lobbying to improve the overall hygiene on the wards. She established a rational system for triaging injured soldiers. As the wounded disembarked they were stripped of their blood-soaked uniforms and their wounds cleansed. To prevent cross contamination between soldiers Nightingale insisted that a fresh clean cloth be used for each soldier, rather than the same cloth for multiple patients. She set up huge boilers to destroy lice and she shamed hospital orderlies to remove buckets of human excrement, and to clean up raw sewage that polluted the wards and to unplug latrine pipes. At her behest new windows capable of opening were installed to air out the wards."

Many of our current health practices, such as isolation of contagious patients, avoidance of cross-contamination, aseptic preparation of food, ventilation of wards, and sanitary disposal of human and medical wastes, stem from the work of Florence Nightingale at Scutari. Her statistical analyses proved that the majority of soldiers in the Crimean War died not of bullets wounds, saber thrusts, or shells but from typhus, dysentery, cholera, and scurvy, all of which are preventable conditions. Nightingale not only gave comfort to the soldiers but also wrote letters to the families of those who died or were dying, and in these letters she explained the circumstances of the illness and death and often she included some of the soldier's personal effects. As such, she was an early practitioner of hospice medicine.

After the war she returned to England, and in 1860 she established the Nursing School for the Training of Nurses—the forerunner of all nursing training. Her wartime experiences convinced her that the health administration of the British Army was in need of reform. She put to good use her knowledge of statistics in assembling data on how administrative inadequacies affected patients' health. Her analysis of mortality showed that even in peacetime, mortality was higher in the military than among civilian males of similar age. She ascribed this higher mortality to contagious disease and emphasized that it could be lowered by improved sanitary conditions. She published a 1,000-page report, *Notes on Matters Affecting the Health, Efficiency and Hospital Administration of the British Army,* and she was the source of inspiration for the more humane and efficient treatment of the wounded in both the American Civil War and the Franco-Prussian War. The stirring record of her deeds led to the founding of the Red Cross Society.

Despite the critics who described Florence Nightingale as a woman "driven by demonic frenzy, a complete egoist, full of a very tiresome religiosity, not very intelligent" and a "megalomaniac who would think of nothing but how to satisfy that singular craving of hers to do something," there remains the positive image of a meticulous record keeper, an able administrator who reformed the terrible conditions that existed in the Crimea, a human being who treated the common soldier with respect rather than brutality, and a compassionate creature who toured the wards alone at night by the light of a Turkish lamp, providing succor to those in need. Her legacy in the field of nursing as well as infection control, hospital epidemiology, and hospice care remains an inspiration to all who seek to improve the public health.

Cure

To treat the violent diarrhea of cholera, health authorities in Germany, England, and France recommended a variety of nostrums and quack remedies: vinegar, camphor, wine, horseradish, mint, mustard plaster, leeches, bloodletting, laudanum, calomel, steam baths, and hot baths. None were effective, and most patients died whether they were treated or not. Since cholera was thought to be due to a miasma, the houses where cholera victims lived (and died) were fumigated using a smoke pot. This, too, did not curb the spread of disease. However, in 1831, fully 50 years before the "germ" of cholera had been identified by Robert Koch, a 22-year-old physician, William O'Shaughnessy, boldly proposed that cholera might be cured by the application of chemistry. After analyzing the blood of a patient with the blue stage of cholera, he suggested that therapy be "to restore its deficient saline matters . . . by absorption, by imbibition, or by injection aqueous fluid into the veins." Dr. Thomas Latta carried out the first practical application of intravenous injections of a salt solution on 15 May 1832 in patients suffering with cholera. All the treated patients improved. He cautioned, "Although by injection of water and salts . . . we may restore the efficient fluids of the body and bring back the blood to its normal state . . . we must still remember that the unknown remote cause, and other agents are still in operation, and require to be remedied before a perfect cure can be performed." This practical treatment saved lives (8 of 25) and was regarded as "the working of a miraculous . . . agent . . ." Indeed, the mortality during a 1906 epidemic was reduced from 70% to 40% by the administration of intravenous hypertonic solutions. However, the therapy fell into disuse when the pandemic subsided, and cathartics and bloodletting prevailed throughout the 19th century. Clearly, the simple and effective therapy of Latta-O'Shaughnessy was far ahead of its time.

It took 160 years for research to show why the therapy of Latta and O'Shaughnessy worked: the cholera toxin acts on the absorptive cells of the intestine, shutting down one major route of sodium transport which allows sodium into the cells together with chloride, but the toxin leaves unaffected the transport system that brings sodium and glucose simultaneously into the cells from the lumen of the intestine. This transport system works only when both sodium and glucose are present, and it remains active during diarrhea. Armed with this knowledge, it was possible (in 1940) to make the appropriate solution with salts and glucose to rehydrate patients suffering from severe cholera-induced diarrhea. Today, oral rehydration treatment (ORT) involves oral or intravenous administration of a solution containing glucose, sodium chloride, potassium, and lactate; the cost is $5 per quart. Another type of ORT, called food-based ORT, substitutes starches and proteins for the glucose. Cereal grains and beans are the source of starches and proteins, and they work effectively to reduce both diarrhea and mortality.

Consequences

It has been claimed that pollution, especially sewage, has ended more lives than smallpox and bubonic plague. Cholera, the most feared of all these sewage-related diseases, has provoked global horror and terror. However, it also changed the way in which people perceived disease, transmission, and the manner by which human health could be preserved. In countries where an adequate sewage and water supply system has been established, cholera has been eradicated. However, it remains endemic in areas of Asia and Africa and in some countries in South and Central America. Today, quarantining and treatment of the sick, along with ensuring that infected individuals do not contaminate drinking water and food, accomplish control of cholera. Provision of potable water almost always reduces epidemic spread.

Cholera is a disease that resulted in the institution of sanitary reforms and the rise of public health, but it has also been used to scapegoat those suspected of being the carriers of disease, especially immigrants and the poor. At times, the public health organizations designed to protect the people at large became the "health police" and those who were already marginalized were labeled a threat to all of society. Sometimes all members of the ethnic or cultural minority to which the sick belonged were stigmatized. Naturally, the burden and blame fell, as it always does, on those without resources or political power—the immigrants and the urban poor. However, the pandemics of cholera also had a positive effect: the fear and horror of cholera promoted the establishment of a public health system within many countries, fostered the nursing profession, and inspired the formation of international bodies to monitor and control the global spread of all infectious diseases.

4

Smallpox: the Speckled Monster

Dateline: The Future
Place: Anywhere, U.S.A.
Compiled by the American Associated Press Wire Services

President N. E. Dee Lay went on television last night to report an outbreak of smallpox in the city of Anywhere. He urged calm and reassured viewers that there was "sufficient vaccine to control its spread." He went on to say that this unprecedented attack on the civilian population in a city of 2.5 million had been contained. It was, he said, the work of an as yet unidentified terrorist group and the "perpetrators would soon be brought to justice." In his 15-min speech to the nation, President Dee Lay recounted what had occurred in the past 2 weeks.

"Twelve days ago, four students who had been attending a talk by Vice President N. Watts Niu at the University of Anywhere appeared at the campus Health Center complaining of fever and muscle aches. After they were diagnosed as having the flu, they were promptly sent to their dorm rooms with aspirin, ibuprofen, and instructions to drink lots of fluid and to stay in bed. A day later, several students broke out in a rash. When the students observed that the rash developed into patches of small pimples, they became alarmed and went to the emergency room of the local hospital. The emergency room physician made an initial diagnosis of chicken pox and took a swab specimen from the oozing pustule; this was sent to the laboratory for examination under the electron microscope. The lab report came back in 24 h and indicated that the structure of the virus was not consistent with chicken pox but more closely resembled the smallpox virus. Over the next 4 days, the clinical picture was similar to the description given by Richard Preston in his book *Demon in the Freezer*: 'The . . . blotches broke out into seas of tiny pimples. They were sharp feeling, not itchy, and by nightfall they covered [the] face, arms, hands and feet. Pimples were rising out of the soles of [the] feet and on the palms of the hands, too. During the night, the pimples developed tiny, blistered heads, and the heads continued to grow larger . . . rising all over the body, at the same speed, like a field of barley sprouting after the rain. They hurt dreadfully, and they were enlarging into boils. They had a waxy, hard look and they seemed unripe . . . fever soared abruptly and began to rage. The rubbing of pajamas on [the] skin felt like a roasting fire. By dawn, the body had become a mass of knob-like blisters. They were everywhere, all over . . . but clustered most thickly on the face and extremities . . . The inside of the mouth and ear canals and sinuses had pustulated . . . it felt as if the skin was pulling off the body, that it would split and rupture. The blisters were hard and dry, and they didn't leak. They were like ball bearings embedded in the skin, with a soft velvety feel on the surface. Each pustule had a dimple in the center. They were pressurized with an opalescent pus. The pustules began to touch one another, and finally they merged into confluent sheets that covered the body, like a cobblestone street. The skin was torn away . . . across . . . the body, and the pustules on the face combined into a bubbled mass filled with fluid until the skin of the face essentially detached from its underlayers and became a bag surrounding the tissues of the head . . . tongue, gums, and hard palate were studded with pustules . . . the mouth dry . . . The virus had stripped the skin off the body, both inside and out, and the pain . . . seemed almost beyond the capacity of human nature to endure.'

The electron microscopy technician who examined the specimen was vaccinated, and to prevent further spread of smallpox the entire hospital was placed under a strict quarantine. Police and National Guard troops were dispatched to surround the hospital, and no one was permitted to enter or leave. Infection control nurses began to interview the sick individuals to find out the identities of the people with whom they had come into contact. Because the existing supply of vaccine in Anywhere was limited, the City Health Department notified the CDC and an additional vaccine supply was provided and used to vaccinate everyone who had been exposed to the infected individuals.

The President went on to say that he regretted that the public had not been informed through normal hospital channels but that this was due to the hospital lockdown. Instead, he said, family members and friends were told of the outbreak through telephone calls from physicians, nurses, the sick, and visitors. This spawned rumors, and the evening late news reported that Anywhere Memorial Hospital had been quarantined because of an infectious disease that could possibly be Ebola, Hong Kong flu, meningitis, measles, or smallpox. "It is true that 2 weeks earlier the FBI had received information on a possible terrorist threat that was to coincide with the Vice President's speech where there would be a thousand attendees in the University auditorium, but since there was no independent confirmation of this, the threat was dismissed as a hoax. I regret that I was unable to notify you earlier and apologize for the confusion and anger it may have caused. Again, there is no reason for panic. We have called in Homeland Security as well as FEMA, the CIA, FBI, and CDC to deal with this outbreak. We expect to have the current situation under complete control in the next two weeks. Good night, and God bless America."

Two days following the President's televised speech, 20,000 residents of the city were vaccinated. One of the students died, and there were 50 new cases of smallpox.

A week after the President's speech, 80,000 people had been vaccinated and 2 children in a neighboring city came down with smallpox. It was unclear whether these cases were due to a new wave of infection or resulted from contact with the first wave of patients. The CDC reported that the smallpox virus was not an engineered variety and was similar to stocks placed in deep freeze in 1989.

Two weeks after the President's speech, 200 probable cases had been reported. The CDC received thousands of requests for vaccine from individual physicians and announced that it would distribute vaccine through state health departments. It cautioned that the quantity of vaccine available can cover only 15% of the residents of the city. To reassure the public, federal officials announced a crash vaccine production program that will yield sufficient vaccine; however, when it became known that this would take 24 months, looting and rioting broke out.

A month after the President's speech, more than 700 cases worldwide had been recorded. The death rate among the infected was 30%. Dramatic footage of young children covered with pox drove thousands of people to emergency rooms and doctors' offices all over the country to evaluate rashes and fevers. The capacity of hospitals to cope with quarantine was exceeded, and suspected cases were isolated in the Armory and Convention Center.

Two months after the President's speech, 15,000 cases and more than 2000 deaths had been reported worldwide. Although the number of new cases appeared to be declining, civil unrest continued in many parts of the world. Some countries refused to admit U.S. citizens without proof of a recent smallpox vaccination. Other countries established a 14-day quarantine on all persons coming from abroad. Domestic and international travel had been severely curtailed. Small businesses in cities where smallpox had been recorded failed because they were unable to get supplies and customers were reluctant to shop. Attendance at sporting events and theatres was greatly reduced. Schools closed because teachers refused to come to work and parents were fearful of exposing their children to "the speckled monster." Smallpox has now spread throughout the world, and human rights organizations have reported cases where smallpox patients were left to die or abandoned or those recovered are denied food and housing. The disease has sown public panic, disrupted and discredited official institutions, and shaken public confidence in the government.

This fictional account is based on the article *Smallpox: An Attack Scenario* by Tara O'Toole, published in *Emerging Infectious Diseases*. This scenario need not come to pass as long as we have a better understanding of the power (and limitations) of vaccination and the protective nature of the immune response.

Smallpox Spreads

Over the centuries, smallpox has killed hundreds of millions of people. In the 20th century alone, it killed at least 300 million people. Smallpox is indiscriminate, with no respect for social class, occupation, or age; it has killed or disfigured princes and paupers, kings and queens, children and adults, farmers and city dwellers, generals and their enemies, rich and poor. It is suspected that humans first acquired the infectious agent from one of the pox-like diseases of domesticated animals in the earliest concentrated agricultural settlements of Asia or Africa when humans began to maintain herds of livestock, some time after 10,000 BC.

From its origins in the dense agricultural valleys of the great rivers in Africa and India, smallpox spread to China. Trade caravans assisted in the spread, but at the time of the birth of Christ it was probably not established in Europe because the populations were too small and were greatly dispersed. In AD 100 there was a catastrophic epidemic of what is suspected to have been smallpox called the Plague of Antonius. The epidemic started in Mesopotamia, and the returning soldiers brought it home to Italy. It raged for 15 years, and there were 2,000 deaths daily in Rome. In the western part of Eurasia the major spread of smallpox occurred in the eighth and ninth centuries during the Islamic expansion across North Africa and into Spain and Portugal. As the disease moved into central Asia, the Huns were infected in either Persia or India. In the fifth century, when the Huns descended into Europe, they may have carried smallpox with them. By AD 1000, smallpox was probably endemic in the more densely populated parts of Eurasia, from Spain to Japan, as well as the African countries bordering the Mediterranean Sea. Assisted by the caravans that crossed the Sahara, smallpox spread into the more densely populated kingdoms of West Africa, and it was repeatedly introduced into the port cities of East Africa by Arab traders and slavers. The movement of people to and from Asia Minor during the Crusades in the 12th and 13th centuries helped reintroduce smallpox to Europe. By the 15th century, smallpox was in Scandinavia; by the 16th century, it was established in all of Europe except for Russia. As Europe's population increased and people were crowded in the cities, smallpox epidemics appeared with increasing intensity and frequency.

As British, French, and Spanish explorers and colonists moved into the newly discovered continent of America, smallpox went with them. In 1521, smallpox allowed Hernan Cortes to topple the Aztec Empire with fewer than 600 men. Cortes and his Spanish conquistadors landed in the Yucatan on the eastern coast of Mexico and quickly moved on to the Aztec capital of Tenochtitlan (now Mexico City), where the outnumbered Spaniards lost one-third of their troops. Forced to retreat, they expected a final and crushing offensive by the Aztecs, one that would result in their complete defeat. The attack never came. On 21 August the Spaniards stormed the city, only to find that a greater force—disease—had ensured their victory. "The houses and stockades . . . were full of heads and corpses. It was the same in the streets and courts . . . one could not walk without treading on the bodies and heads of the dead Indians . . . the stench was so bad that no one could endure it." Smallpox had already arrived in Mexico: in 1520, an expedition arrived from Spanish Cuba, and among the crew was a smallpox-infected slave. From this initial infection, smallpox spread from village to village throughout the Yucatan. So great was the die-off that there were not enough people to farm the land or protect the cities. Famine and havoc resulted. The paralyzing effect of the epidemic explains why the Aztecs did not pursue the demoralized Spaniards. Instead, the Spaniards were able to rest and regroup and, in time, gathered Indian allies. Smallpox was a devastatingly selective disease—only the Aztecs died, and the Spaniards were left unharmed—and it so demoralized the survivors that the Amerindians had no doubt that the Spaniards were the beneficiaries of divine favor. It was this perceived superior power of the God worshipped by the Spaniards that led the Aztecs and other Amerindians to accept Christianity. Without further resistance they submitted to Spanish control. In the time of Cortes and his conquistadors, three million Amerindians, an estimated one-third of the total population of Mexico, were killed by smallpox.

Twelve years later (1532) another Spanish conquistador, Francisco Pizarro, led a group of 168 soldiers and captured Atahualpa, the leader of the Inca Empire. Pizarro held Atahualpa prisoner for 8 months, demanded a ransom of gold from his subjects for a promise of freedom, and, after the gold was delivered, reneged on the promise and had Atahualpa executed. Again, smallpox had contributed to Spanish success in conquering South America. Smallpox had arrived in Peru by land in 1526, killing much of the Inca population. When Pizarro landed on the coast of Peru in 1531, the stage for the Inca conquest had already been set; the smallpox epidemic that had preceded the conquistadors had weakened the Inca Empire, there was civil war, and Atahualpa's army was vulnerable and in disarray. Victory for Pizzaro's conquistadors had been ensured by smallpox.

Smallpox traveled to West Africa with the caravans that moved from North Africa to the Guinea Coast. In 1490, it was spread by the Portuguese into the more southerly regions of West Africa. It was first introduced into South Africa in 1713 in the port of Capetown by a ship carrying contaminated bed linen from India. The first outbreak of smallpox in the Americas was among African slaves on the island of Hispaniola in 1518. Here and elsewhere, the Amerindian population was so decimated that by the time of Pizzaro's conquest around 200,000 had died. With so much of the Amerindian labor force lost to disease, there was an increased need for replacements to do the backbreaking work in the mines and plantations of the West Indies, the Dominican Republic, and Cuba. This need for human beasts of burden stimulated, in part, the slave trade from West Africa to the Americas.

Smallpox is reputed to be the first germ warfare agent. In the war of 1763 between England and France for control of North America, under orders of Sir Geoffrey Amherst (Commander-in-Chief of North America) the British troops were asked, "Could it not be contrived to send the smallpox among those disaffected tribes of Indians? We must on this occasion, use every stratagem in our power to reduce them." The ranking British officer for the Pennsylvania frontier, Colonel Henry Bouquet, wrote back, "I will try to inoculate the Indians with some blankets that may fall into their hands and take care not to get the disease myself." Blankets were deliberately contaminated with scabby material from the smallpox pustules and then delivered to the Indians, allowing the spread of the disease among these highly susceptible individuals. This led to extensive numbers of Indian deaths and ensured their defeat.

Cause

The cause of smallpox is a virus, one of the largest of the viruses; with proper illumination it can actually be seen under a light microscope. However, much of its detailed structure can be visualized only by using the electron microscope. The outer surface (capsid) of the smallpox virus resembles the facets of a diamond, and its inner, dumbbell-shaped core contains the genetic material, double-stranded DNA. The virus has about 200 genes, 35 of which are thought to be involved in virulence.

Most commonly the smallpox virus enters the body through droplet infection by inhalation. However, it can be transmitted by direct contact or through contaminated fomites (inanimate objects) such as clothing, bedding, blankets, and dust. The infectious material from the pustules can remain infectious for months. The virus multiplies in the mucous membranes of the mouth and nose. During the first week of infection there is no sign of illness; however, the virus can be spread by coughing or by nasal mucus at this time. The virus moves on to the lymph nodes and then to the internal organs via the bloodstream. Once there, it multiplies again. It then reenters the bloodstream. Around the ninth day the first symptoms appear: headache, fever, chills, nausea, muscle ache, and sometimes convulsions. The person feels quite ill. A few days later a characteristic rash appears. The individual is infectious a day before the rash appears and until all the scabs have fallen off. Many infected persons die a few days or a week after the rash appears. Destruction of the sebaceous (oil) glands of the skin leaves permanent craterlike scars in the skin, known as pockmarks.

There are two varieties of the smallpox (variola) virus: major and minor. They can be distinguished by differences in their genes. Variola major, the deadlier form, frequently killed up to 25% of its victims, although in naïve populations such as the Amerindians the fatality rate could exceed 50%. Variola minor, a milder pathogen, had a fatality rate of about 2% and was more common in Europe until the 17th century, when it mutated to the more lethal form, perhaps as a result of reintroduction from the Spanish colonies in the Americas. In the 17th century, smallpox was Europe's most common and devastating disease, killing an estimated 400,000 persons each year. It caused one-third of all cases of blindness. By the beginning of the 18th century, nearly 10% of the world's population had been killed, crippled, or disfigured by smallpox. The 18th century in Europe has been called the "age of powder and patches" because pockmarks were so common. The "beauty patch" (a bit of colored material) seen in so many portraits was designed to hide skin scars. In this way, the pockmark set a fashion.

"Catching" Smallpox

Smallpox is a contagious disease of civilization that is spread from person to person without animal reservoirs, i.e., it is not a zoonotic disease. It can exist in a community only as long as there are susceptible humans. It has been estimated that a minimum population of the order of 100,000 persons is needed to ensure that there are enough susceptible persons born annually to sustain the chain of infection indefinitely. Smallpox spreads more rapidly during the winter in temperate climates and during the dry season in the tropics.

Smallpox caused large epidemics when it first arrived in a virgin community where everyone was susceptible, but subsequently most individuals who were susceptible either recovered and were immune or died; then the disease died down. If the infection was reintroduced later, after a new crop of susceptible individuals had been born or migrated into the area, another epidemic would occur. However, after a time the disease would reappear so frequently that only newborns would be susceptible to infection, older individuals being immune from previous exposure. In this situation, smallpox became a disease of childhood similar to measles, chickenpox, mumps, whooping cough, and diphtheria. Sometimes it did not disappear altogether, especially in larger communities, but remained at a low level. However, the infection would break out as an epidemic every 5 to 15 years when enough susceptible individuals had accumulated.

Control by Variolation, Eradication by Vaccination

Today, prevention of smallpox is often associated historically with Edward Jenner, but even before Jenner's scheme of vaccination, techniques were developed to induce a mild smallpox infection. The educated classes called this practice of folk medicine "inoculation" from the Latin "inoculare," meaning "to graft," since it was induced by cutting. It was also called "variolation" since the scholarly name for smallpox was "variola" from the Latin "varus" meaning "pimple." There were numerous techniques of variolation. The Chinese avoided contact with the smallpox-infected individual, preferring that the person either inhale a powder from the dried scabs shed by a recovering patient or place the powdered scabs on a cotton swab and insert this into the nostrils. In the Near East and Africa, material from the smallpox pustule was rubbed into a cut or a scratch in the skin.

Variolation came to Europe through the efforts of Lady Mary Wortley Montagu, a highborn English beauty. In London in 1715, at age 26, she contracted smallpox. She recovered, but smallpox-induced facial scars and the loss of her eyelashes marred her beauty forever. Her 22-year-old brother was less fortunate, however, and died of smallpox. The following year, her husband was appointed ambassador to Turkey, and she accompanied him. There she learned of the Turkish practice of variolation. She was so impressed by what she saw that she had her son inoculated.

On returning to England in 1718, she propagandized the practice, and she had her daughter inoculated against smallpox at age 4. She convinced Caroline the Princess of Wales (later Queen Consort to George II) that her children should be variolated, but before that could happen the royal family required a demonstration of its efficacy. In 1721 the "Royal Experiment" began with six prisoners who had been condemned to death by hanging and were promised their freedom if they subjected themselves to variolation. They were inoculated, suffered no ill effects, and, upon recovery, were released as promised. Despite this, the Princess wanted more proof and insisted that all the orphan population of St. James Parish be inoculated. In the end only six children were variolated, and with great success. The Princess, now convinced, had two of her children (Amelia and Caroline) variolated. This too was successful. However, Lady Mary met with sustained and not entirely inappropriate opposition to the practice of variolation. The clergy denounced it as an act against God's will, and physicians cautioned that the practice could be dangerous to those inoculated as well as others with whom they came in contact. The danger of contagion was clearly demonstrated when the physician who had previously inoculated Lady Mary's children variolated a young girl, who then proceeded to infect six of the servants in the house; although all the infected persons recovered, it did cause a local epidemic.

Variolation came into practice quite independently in the English colonies in America. Cotton Mather, a Boston clergyman and scholar, was a fellow of the Royal Society of London. Although he had read the account of Emanuel Timoni, a physician living in Turkey, he wrote to the Society that he had already learned of the process of variolation from one of his African slaves, Onesimus. In April 1721 there was a smallpox outbreak in Boston. Half the inhabitants fell ill, and the mortality rate was 15%. During this time, Mather tried to encourage the physicians in the city to carry out variolation of the population. The response from all the physicians but one, Zabdiel Boylston, was negative. On 26 June, Boylston variolated his 6-year-old son, one of his slaves, and the slave's 3-year-old son without complications. By 1722, Mather, in collaboration with Boylston, had inoculated 242 Bostonians and found that it protected them. His data showed a mortality rate of 2.5% in those variolated compared to the 15 to 20% death rate during epidemics. Boylston's principal opponents, however, were his fellow physicians, not the clergy.

On 8 May 1980, smallpox was certified as eradicated by the World Health Assembly, but eradication did not result from variolation. It was eradicated thanks to the development of the first vaccine by Edward Jenner (1749 to 1823), an English physician. Jenner did not know what caused smallpox or how the immune system worked; nevertheless, he was able to devise a practical and effective method for immune protection against attack by the lethal virus. Essentially he took advantage of a local folktale and turned it into a reliable and practical device against the ravages of smallpox.

The farmers of Gloucestershire observed that if a person contracted cowpox—a disease characterized in cows by blisters on the udders that clear up quickly—they were immune to smallpox. In 1774 a cattle breeder named Benjamin Jesty contracted cowpox from his herd, thereby immunizing himself against smallpox. He then deliberately inoculated his wife and two children with cowpox. They remained immune even 15 years later when they were deliberately exposed to smallpox. Jenner, however, not Jesty, is given credit for vaccination because he carried out experiments in a systematic manner over a period of 25 years to test the farmer's tale. Jenner wrote, "It is necessary to observe, that the utmost care was taken to ascertain, with most scrupulous precision, that no one whose case is here adduced had gone through smallpox previous to these attempts to produce the disease. Had these experiments been conducted in a large city, or in a populous neighborhood, some doubts might have been entertained; but here where the population is thin, and where such an event as a person's having had the smallpox is always faithfully recorded, no risk of inaccuracy in this particular case can arise." As with the earlier "Royal Experiment," Jenner felt free to carry out human experimentation without any reservations, although it should be noted that all of his patients wanted the vaccination for themselves or their children. On 14 May 1796 he took a small drop of fluid from a pustule on the wrist of Sarah Nelms, a milkmaid who had an active case of cowpox, and smeared it onto the unbroken skin of a small boy, James Phipps. Six weeks later, Jenner tested the ability of his "vaccine" (from the Latin "vacca" meaning "cow") to protect against smallpox by deliberately inoculating the boy with material from a smallpox pustule. The boy showed no reaction—he was immune to smallpox. In the years that followed, "poor Phipps" (as Jenner called him) was tested for immunity to smallpox about a dozen times, but he never contracted the disease. Jenner wrote up his findings and submitted them to the Royal Society; however, to his great disappointment, his manuscript was rejected. It has been assumed that the reason was that he was simply a country doctor and not a part of the scientific establishment. In 1798, after a few more years of testing, he published a 70-page pamphlet, *An Inquiry into the Causes and Effects of Variolae Vaccinae,* in which he reported that the inoculation with cowpox produced a mild form of smallpox that would protect against severe smallpox, as did variolation. He correctly observed that the disease produced by vaccination would be so mild that the infected individual would not be a source of infection to others, a discovery of immense significance.

The reactions to Jenner's pamphlet were slow, and many physicians rejected his ideas. However, within several years some highly respected physicians began to use what they called the "Jennerian technique" with great success. By the turn of the century, the advantages of vaccination over variolation were clear and Jenner became famous. In 1802, Parliament awarded him a prize of £10,000, and in 1807 it awarded another £20,000. Napoleon had a medal struck in his honor, and he received honors from governments around the world. By the end of 1801, his methods were coming into worldwide use, and it became more and more difficult to supply sufficient cowpox lymph or to ensure its potency when shipped over long distances. To meet this need Great Britain instituted an Animal Vaccination Establishment in which calves were deliberately infected with cowpox and lymph was collected. Initially this lymph was of variable quality, but it was found that the addition of glycerin prolonged preservation; this then became the standard method of production, and the first "glycerinated calf's lymph" was distributed in 1895. With general acceptance of Jenner's findings, the hide of the cow called Blossom, which Jenner had used as the source of cowpox in his experiments, was enclosed in a glass case and placed on the wall of the library in St. Georges's Hospital in London, where it remains to this day.

How vaccination works

Why does recovery from a smallpox infection, variolation, or vaccination work to protect against this disease? The simple explanation is that the individual has become immune. Being immune means that the body is able to react specifically to a foreign material as a result of previous exposure. To be effective, the immune response has to be swift the second time around. So what are the essential properties of our immune system that protect against a foreign invader? First, it must be able to distinguish foreign substances—"not self"—from "self." Second, it must be able to remember a previous encounter; that is, there must be memory. Third, it must be economical; that is, it should be turned on and off as needed. And fourth, it must be specific for the foreign substance.

In 1890 two critical discoveries related to immunity were made: Emil von Behring (1854 to 1917) and Shibasaburo Kitasato (1852 to 1931), both working with Robert Koch in Berlin, found that if small amounts of tetanus toxin were injected into rabbits, the rabbit serum contained a substance that would protect another animal from a subsequent lethal dose of toxin. They called the serum that was able to neutralize the toxin "immune serum." In effect, by toxin inoculation, the rabbit had been protected against tetanus. Further, the immune protection seen by von Behring and Kitasato could be transferred from one animal to another by injection of this "immune serum." In other words it was possible to have passive immunity; that is, immunity could be borrowed from another animal that had been immunized with the active foreign substance when its serum was transferred (by injection) to a nonimmunized animal. (Passive immunization, or "immunity on loan," can be lifesaving in the case of toxins such as bee, spider, or snake venoms since it is possible to counteract the deadly effects of the poison by injection of serum containing the appropriate antitoxin.) The implications of the research on "immune serums" for treating human disease were quite obvious to von Behring and others in the Berlin laboratory. Without waiting for the identification of the active ingredient in "immune serum," the German government began to support the construction of factories to produce different kinds of "immune sera" against a wide range of toxins.

von Behring and Kitasato gave the name "antitoxin" (meaning "against the poison") to the substance in immune serum that neutralized these toxins. An antitoxin is one of many kinds of active materials found in immune serum after a foreign substance is injected into rabbits, horses, chickens, mice, guinea pigs, monkeys, rats, or humans; the general term for the active material is "antibody." Simply put, antitoxin is one kind of antibody.

After Paul Ehrlich (1854 to 1915) joined Koch's laboratory in Berlin, he began research on immunity. Ehrlich established the principle that there was a direct (one-to-one) relationship between the amount of toxin neutralized and the amount of antitoxin used. The potency or strength of an antibody is called its titer and is reflected by the degree of dilution that must be made to have a specific amount of antigen that will bind to an equal amount of antibody. (The higher the titer, the greater the potency of the serum; put another way, a high-titer immune serum can be diluted to a greater degree to achieve the same neutralizing effect as would a weaker immune serum.) Not only was Ehrlich's titer principle of theoretical interest, but also it enabled von Behring's antitoxin to be provided in standardized amounts and strengths. Ehrlich also found that although the diphtheria and tetanus toxins lost their toxic capacity with storage, they still retained their ability to induce immune serum.

Ehrlich went on to study other toxins and found that the serum made against one toxin did not protect against another toxin; i.e., the antitoxin made by the animal was specific. Since this specificity was true for almost any foreign substance, he concluded that antibodies were specific for the foreign material against which they were produced. Substances such as a toxin, that were antibody generators, were called antigens. Antigens are defined as any material that is foreign to the body, be it a bacterium or its toxin, a virus such as smallpox, a piece of tissue from another individual, or a foreign chemical substance such as a protein, nucleic acid, or carbohydrate polymer (polysaccharide).

Ehrlich also formulated a theory of how antigen and antibody interact with one another. It was called the "lock-and-key" theory, in which the tumblers of the lock were the antigen and the teeth of the key were the antibody. One specific key would work to open only one particular lock. Another way of looking at the interaction of antigen with antibody is the way a tailor-made glove is fitted to the hand. One such glove will fit only one specifically shaped hand. Similarly, only when the fit is perfect can the antigen combine with the antibody. Although antigens are large molecules, only a small surface region is needed to determine the binding of an antibody to it. These small antigenic determinants are called epitopes.

Vaccination against smallpox results in protection because the antibodies produced against Variola vaccinae are able to neutralize the variola major virus. Simply stated, it is because the epitopes of the cowpox antigens are so similar to those of smallpox that when the body produces antibodies to cowpox antigens, these antibodies are also able to fit exactly to the smallpox epitopes (i.e., the antibodies are said to be cross-reactive) and variola major is prevented from multiplying.

How immunity works

Blood consists of a salty protein-containing fluid, the plasma, in which float red blood cells and white blood cells. (When blood plasma clots, the remaining fluid, which is deficient in clotting molecules, is called serum.)

The red cells are not involved in immunity, whereas the white cells—lymphocytes, monocytes, and macrophages—are critical components of the immune system. Antibodies are made by a particular kind of lymphocyte called the B lymphocyte or B cell. On the surface of the B lymphocyte is an antibody molecule that serves as a receptor for the foreign antigen. Once bound to the B-cell surface, the B cell (now called a plasmablast) is triggered to divide, and the population of identical offspring from such division are clones. The members of the clone population are called plasma cells. Division to produce an adequate number of plasma cells may take several days since the division rate is about 10 h per division. Plasma cells are cellular factories that can manufacture large quantities of antigen-specific antibody, which they secrete into the blood plasma. A single plasma cell can secrete over 2,000 molecules of antibody/s or about 10^6 molecules/h, and it is possible for different plasma cells to synthesize 10^{15} *different* kinds of antibodies, each having a *different* specificity for the antigen that triggered division in the first place. The specific antibody produced can protect several ways: it can neutralize the antigen, it can aggregate (clump) free virus (or bacteria) in the blood, it may cause virus-infected cells (or bacteria) to lyse (disintegrate), it may block virus entry into cells, or it may make the microbes more attractive so that phagocytes readily ingest them. Not all of the B cells differentiate into plasma cells; some remain quiescent and persist as memory cells. When the immune system encounters the same antigen at a later time, memory cells undergo rapid division, plasma cells are produced, and antibody is synthesized and released. This rapid response is due to memory of a past antigenic encounter and is the basis of immunization with booster shots.

Antibody, produced by B cells, is one part of the immune system involved in protective immunity. Another part of the immune system involves a different set of lymphocytes, the T-cell variety, and is concerned with non-antibody immunity. In the 1960s it was discovered that the bone marrow and thymus are the "master organs" of the mammalian immune system. The bone marrow is the site where all the different cell types of the blood and the immune system are made, and it is here that the blood stem cells also exist. Stem cells are primitive, unspecialized cells, and they have no function other than to divide and to make other cells. However, when a stem cell divides to produce two daughter cells, the offspring are unalike: one daughter cell is an identical replica of the stem cell, but the other can grow and become a very specialized cell. In this way, stem cells are self-renewing and also give rise to a variety of specialized cell types, including B and T lymphocytes. A T lymphocyte arises in the bone marrow but migrates via the bloodstream from the bone marrow to a gland in the neck called the thymus.

The thymus is made of lymphoid tissue and is located just above the heart. In children the gland is rather large; as we age, it tends to atrophy and shrink. The thymus is the place where the T lymphocytes from the bone marrow establish residence for some time and where they mature (or become "educated") and multiply. It is because of their maturation in the thymus that they are called "thymus-educated" (T) cells. While in the thymus, some T cells are selected to become either helper T cells (also called CD4 cells) or killer T cells (also called CD8 cells). The killer cells are also referred to as "cytotoxic T lymphocytes" (CTLs). Some of the offspring of these T cells leave the thymus and, by way of the bloodstream, reach and settle in other lymphoid organs (the spleen, the tonsils, and the variously distributed lymph nodes) or circulate in the blood.

Antibody cannot attack microbes—viruses and rickettsias—that are concealed within cells. That is the job of the CTLs. The CTLs patrol the body in search of cells compromised by such internal pathogens. Because they are capable of attacking and destroying any cell in the body that appears to be "not self," they are aptly called killers. However, the killer cell does not direct its action against the virus itself but, rather, against the cell (no longer recognized as self) in which the virus is hiding. Killing may occur in one of two ways. In the first mechanism, when a foreign cell is encountered, the CTLs, which have CD8 on their surface, attach to it and punch in a "security code" that then triggers a self-destruct mechanism in the non-self cell. In effect, the CTL need not shoot the invader with a "silver bullet" but instead simply tells the pathogen-infected cell, "die." The other mechanism is for the CTL to punch a hole in the infected cell by using a pore-forming molecule and then inject the infected cell with a lethal cocktail of destructive enzymes. The CTL response is strong within 5 days after exposure to foreign antigen, peaks between days 7 and 10, and then declines over time. (Antibody formation usually follows the CTL response.)

How do the T cells act to defend against infection? The first line of defense in cell-mediated immunity is the macrophage. Macrophages, the "foot soldiers" of the immune system, are also the cellular vacuum cleaners or filter feeders of the immune system. They ingest, digest, and then regurgitate the foreign pathogens (virus or bacterium) or even the dead and dying pathogen-containing cells they have eaten. In this way, they are able to display the broken antigenic bits on their outer surface in association with another molecule called the major histocompatibility complex (MHC) II. This combination of MHC II and foreign antigen provides information to a helper T lymphocyte. The helper T cell has the CD4 receptor (this receptor is not an antibody), and once the CD4 binds to the MHC-antigen combination and receives the information, it is triggered to divide, forming a clone. However, instead of secreting antibody, it secretes chemical messengers called lymphokines, which activate other T lymphocytes or macrophages or can activate the B cells which also carry MHC II so that they begin the process of antibody synthesis and secretion. In addition, as with B cells, memory T cells are left after an encounter with a foreign antigen so that the response to a second encounter is swift and specific. The CD4 helper T cell is the keystone that maintains the connection between the two branches of the immune system: cell-mediated immunity and antibody-mediated immunity. Antibodies prevent foreign invaders from infecting cells, whereas the members of the cell-mediated branch search out and destroy infected or foreign cells.

CTLs do not destroy the body's own cells or the helper T cells because all the cells in the body (except for red blood cells) express another histocompatibility molecule, MHC I. This allows infected cells to signal their plight to the CTL, which is also MHC I. Once this information is received by the CTL, the infected cell is killed. However, because the helper T cell is MHC II, either it is not recognized by the CTL or the information received tells the CTL that the helper T cell is self and killing does not occur.

Attenuation and immunization

In 1875, Louis Pasteur attempted to induce immunity to cholera. He was able to grow the cholera bacteria that killed chickens in the laboratory and to reproduce the disease after injecting healthy chickens with these bacteria. The story is told that he placed his cultures of cholera on a shelf in his laboratory, exposed to the air and went on summer vacation; when he returned to his laboratory, he injected chickens with the old culture and the chickens became ill but, to Pasteur's surprise, did not die. Pasteur then grew a fresh culture of the bacteria with the intention of injecting another batch of chickens, but the story goes that he was low on chickens and used those he had previously injected. Again, he was surprised to find that the chickens did not die of cholera but were apparently protected from the disease. He recognized that aging and exposure to air (and possibly the heat of summer) had lowered the virulence of the bacteria and that such an attenuated or weakened strain might be used to protect against disease. In honor of Jenner's work a century earlier, he called the attenuated strain a vaccine. Pasteur extended these findings to other diseases and in 1881 was able to protect sheep against anthrax. In 1885 he administered the first vaccine (prepared from dried spinal cords of rabies-infected animals) to a human, a young boy named Joseph Meister, who had been bitten by a rabid dog. The boy lived, and Pasteur had proved that vaccination worked. Attenuation was also used in the development of BCG (bacillus Calmette-Guérin), the vaccine for tuberculosis (see p. 126); it took 231 subcultures of a bovine tubercle bacillus in an ox bile-containing medium every 3 weeks over a period of 13 years before the bacterium was attenuated. However, some see a problem with attenuation: how can it be ensured that the benign form will not, in time, revert to its virulent state and, instead of giving protection, cause disease?

Consequences

It is difficult to underestimate the contribution of immunization to our well-being. It has been estimated that, were it not for childhood vaccinations against diphtheria, pertussis, measles, mumps, smallpox, and rubella, as well as protection afforded by vaccines against tetanus, cholera, yellow fever, polio, influenza, hepatitis B, bacterial pneumonia, and rabies, childhood death rates would probably hover in the range of 20 to 50%. Indeed, in countries where vaccination is not practiced, the death rates among infants and young children remain at that level.

After Pasteur's success in creating standardized and reproducible vaccines at will, the next major step came in 1886 from the United States. Theobald Smith and Edmund Salmon published a report on their development of a heat-killed cholera vaccine that immunized and protected pigeons. Two years later, these two investigators claimed priority for having prepared the first killed vaccine. Although their work appeared in print 16 months before a publication by Chamberland and Roux (working in the laboratory of Louis Pasteur) and with identical results, the fame and prestige of Pasteur were so widespread that the claim of "first killed vaccine" by Smith and Salmon was lost in the aura accorded those who worked in Pasteur's Institute in Paris. In the period from 1870 to 1890, killed vaccines for typhoid fever, plague, and cholera were developed, and by the turn of the century there were two human attenuated virus vaccines (*Vaccinia* and rabies) and three killed ones. Although there were other successes, such as a formalin-inactivated diphtheria toxin (toxoid) in 1923 and BCG in 1927, the golden age of vaccine development really began in 1949, when viruses could be grown in chicken embryos and in human cell cultures. This led to the first live polio vaccine (1950); the Salk formalin-inactivated polio vaccine; attenuated Japanese encephalitis, measles, and rubella vaccines; and an inactivated rabies vaccine. In the 1970s and 1980s, bacterial vaccines were made from the capsular polysaccharides of pneumococcus, meningococcus, and *Haemophilus influenzae*. The majority of modern vaccines are being developed via new technologies: DNA vaccines, subunit vaccines (purified proteins or polysaccharides), as well as genetically engineered and virus-transmissible antigens. The older methods continue to be useful, however, as is the case with the live flu virus vaccine. A prominent vaccinologist said, "If these new technologies succeed, the golden age will turn to platinum." To think, all of this began when a perceptive English country doctor, Edward Jenner, turned a folk belief into a practical method to protect humans from their own diseases.

5

Bubonic Plague

> Towards the beginning of spring . . . the doleful effects of the pestilence began to be horribly apparent by symptoms . . . an issue of blood from the nose was a manifest sign of inevitable death; but in men and women alike it first betrayed itself by the emergence of certain tumors in the groin or armpits, some of which grew as large as a common apple, others as an egg . . . from the two said parts of the body it soon began to propagate and spread itself in all directions; after which the form changed, black spots . . . making their appearance in many cases on the arm or thigh or elsewhere, now few and large, then minute and numerous almost all within three days from the appearance of said symptoms, sooner or later, died . . . despite all that human wisdom and forethought could devise to avert it, as the cleansing of the city from many impurities by officials appointed for the purpose, the refusal of entrance of all sick folk, and the adoption of many precautions for the preservation of health; despite all humble supplications addressed to God, and often repeated both in public procession and otherwise, by the devout; towards the beginning of spring of the said year the doleful effects of the pestilence began to be horribly apparent by symptoms that showed as if miraculous . . . such as were left alive inclining almost all of them to shun and abhor all contact with the sick and all that belonged to them, thinking thereby to make their own health secure . . . they banded together and formed communities in houses where there were no sick, and lived a separate and secluded life, which they regulated with the utmost care . . . eating and drinking moderately of the finest wines, holding converse with none but one another . . . diverting their minds with music and, other delights as they could devise.

Giovanni Boccaccio (1313 to 1375), in his Introduction to *The Decameron* (ca. 1350), clearly described the time of the plague in Florence, Italy; he survived this "pestilence"; however, others were not so lucky, and by the time it had subsided the population of Europe and the Middle East had been reduced from 100 million to 80 million people. The pestilence

Boccaccio wrote about and that swept through Europe, the Near East, and Africa from 1346 to 1352 was probably one of the greatest public health disasters in recorded history. It put an end to the rise in the human population that had begun in 5000 BC, and it would take more than 150 years for Europe's population to return to its former size. Some believe this crash in population to be Malthus' prophecy (outlined on p. 25) come true, but the historian David Herlihy contends that this outbreak of plague, also called the Black Death, was not a catastrophe promoted by "positive checks" (i.e., disease, war, and famine) but an exogenous factor that served to break a Malthusian stalemate. That is, despite fluctuations in population size, relatively stable population levels were maintained over prolonged periods due to "preventive checks" (i.e., changes in inheritance practices, delay in the age of marriage, and birth control). The Black Death did more than break the Malthusian stalemate; it provoked Europeans to restructure their society along very different paths and to institute public health measures to control the spread of this disease.

The Church placed the responsibility for plague on God, suggesting that this was Judgment Day and that nothing could be done to reduce the suffering of sinners. However, the high mortality rate among priests who gave last rites led to a loss of faith in the clergy because they seemed so powerless to prevent death or the spread of death from disease. With the decline in the ability of the church to provide comfort, there was an expanded search for patron saints who not only knew suffering but also had the power to heal the sick. One of these plague saints was St. Roch, and another was St. Sebastian.

Medicine and education were also affected. When the holders of the Chairs of Medicine at the great European universities died, the newer and younger appointees could move into other clinical areas. Surgeons, who wore a costume with a beak containing perfume or spices, a cloak of waxy leather, eye lenses, and a wand with incense, and who cared for the sick, died at higher rates than did the other medical practitioners. Because of this, their role in curing disease was little valued. New prestige fell to the barbers, and bloodletting and surgery became an integral part of their practice rather than barbering alone. This also led to an emphasis on studies of human anatomy in health and disease, and Galen's philosophy, which had no clear theory of contagion, declined in importance.

As the death toll increased, the numbers of learned individuals decreased. This affected particular universities, where lawyers, physicians, and clerics were trained. Restrictions on travel prevented students from enrolling at distant universities, and so local universities were established; it no longer became necessary to travel to Bologna or Paris for an education. This diminished the dominance of certain centers of learning and led to curricular reform, and instruction began to be carried out in the vernacular tongue.

The feudal system of serfs and landlords declined as the practice of monetary payment for manual services took hold. The smaller labor force in the cities meant that laborers had a stronger negotiating position and so could command higher wages; as a consequence, there was an improvement in the standard of living. Another consequence of the Black Death was the employment of economies of scale in oceangoing transport vessels. Bigger ships with smaller crews could remain at sea for longer periods and would be able to sail directly from port to port, but this would require better ship construction, improvements in navigational instruments, and new business enterprises such as maritime insurance to protect the investment in cargo and the ship. As a result, merchants such as bankers and craftsmen became more powerful. The new economy became more diversified, there was a more intensive use of capital, technological innovations became more important, and there was a greater redistribution of wealth.

Cause

As plague raged through medieval Europe, it became increasingly apparent that the disease was contagious. But was it due to the inhalation of the malignant, foul-smelling air (a miasma) or did it result from the drinking of polluted water? One of the first physicians to advance a plausible theory about the possible agent for the plague was Girolamo Fracastoro (1478 to 1533) in his book *On Contagion and Contagious Diseases*. Fracastoro wrote that plague (and other diseases) could be transmitted by "seeds of contagion" or "seminaria." His theory was put into practice when, as physician to Pope Paul III, he recommended the transfer of the Council of Trent from Trent to Bologna as a response to an outbreak of plague. Other physicians of the time, however, did not subscribe to Fracastoro's theory, and soon the practice was displaced by misguided suggestions such as purging, bleeding, fumigating, bathing in urine, and prayer. However, even if the intellectual climate of the time had favored Fracastoro's seeds of contagion, there would have been few ways to precisely characterize the responsible agent. Identifying the 'seed' and preventing its transmission would require both technological innovation and a change in the concept of infectious (contagious) diseases. The novel concept was that disease could result from the invasion of the body by microscopic parasites called germs, and the technological innovation, the microscope, would allow the visualization of Fracastoro's seminaria. However, it would take more than five centuries after the appearance of the Black Death for this nexus of thought and technology to occur. In the middle of the 19th century, two schools, one in France under the leadership of Louis Pasteur (1822 to 1895) and the other in Germany, led by Robert Koch (1843 to 1910), were responsible for assigning a specific microbe to be the cause of a particular disease, i.e., for formulating the germ theory of disease.

In the 1850s, Koch, Pasteur, and their microbe-hunting cohorts, armed with a disease-causing theory as well as microscopes with improved powers of magnification, began searching for agents that might be the cause of plague epidemics. Their chance came in the 1860s, when the disease broke out in the war-torn Yunnan region of China. It then spread to the southern coast of China and, assisted by modern steamships and railways, quickly moved on to the rest of the world. When it arrived in Hong Kong in 1894, Pasteur dispatched Alexandre Yersin, a Swiss-born member of the French medical colonial corps, to isolate the germ of plague. Yersin arrived in Hong Kong when the epidemic was in full force. At first he was not permitted access to the morgue, but through connivance and bribery he was finally able to arrange a visit to examine a sailor who had just died of plague. He punctured the "tumor" on the dead soldier's thigh with a sterile needle and removed some fluid; he then examined the fluid under the microscope, inoculated a few guinea pigs with some of the fluid, and sent the remainder to the Pasteur Institute in Paris. On 24 June 1894, he wrote Pasteur that the fluid was full of bacteria that stained poorly with Gram's stain; i.e., they were gram negative. (Gram stain consists of a tincture of crystal violet. After bacteria are killed and washed, some lose the dye whereas others retain the purple color; the former are called gram-negative and the latter gram-positive bacteria.) Yersin also wrote, "Without question this is the microbe of plague." A few days later, he found that the guinea pigs he had injected with the fluid aspirated from the sailor's tumor (called a bubo) had died, and their bodies swarmed with the same bacteria. Yersin was intrigued by the large number of dead rats in the streets of Hong Kong, as well as in the hospital corridors and the morgue. When he examined some of these dead rats, he found the same bacteria to be present. He correctly concluded that plague infects both rats and humans. At about the same time, Koch, also convinced that a microbe caused plague, sent his associate Shibasaburo Kitasato and a large number of assistants, as well as abundant equipment, to find the plague germ. Kitasato was able to grow bacteria from the finger of a sailor who had died of plague, but the strain was Gram positive. Further, Kitasato was unable to confirm that this microbe could produce plague in humans or other animals. The definitive diagnosis: plague was caused by a rod-shaped, gram-negative bacterium named *Yersinia pestis* after its discoverer Alexandre Yersin (although the name used before 1870 was *Pasteurella pestis*).

Yersin discovered the plague "germ" but did not find how the disease could be transmitted to rats and humans. He was interested in whether the disease could be due to foul-smelling air and what role rats played. The answers to these questions came from Paul-Louis Simond (1858 to 1947), a French army physician sent by Pasteur to Vietnam and India to follow up on Yersin's observations. Simond noted that not only were there large numbers of dead and dying rats in the streets and buildings, but also 20 laborers in a wool factory who had been cleaning the floor of dead rats had died of plague, although none of the other factory workers, who had no contact with rats, became ill. He began to suspect that there must be an intermediary between a dead rat and a human. He noticed that healthy rats groomed themselves and therefore harbored few fleas whereas sick rats were unable to groom their fur and harbored many fleas; when the rats died, the fleas moved off onto other healthy rats or humans. Simond then asked whether the intermediary could be a blood-sucking flea. To test his idea, he did an experiment: he placed a sick rat at the bottom of a jar, and above this he suspended a healthy rat in a wire mesh cage; although the healthy rat had no direct contact with the plague-infected one, it did become infected by exposure to fleas from the sick rat (which he determined could jump as high as 4 in. without any difficulty). As a control, Simond placed a sick rat without fleas together with healthy rats in a jar; none of the healthy ones became sick, but when he introduced fleas into the jar they developed plague and died. On 2 June 1898 he wrote Pasteur that the problem of plague transmission had been solved.

Control by Quarantine

More than 2,000 years ago—long before the germ theory and powerful microscopes existed—the physicians Hippocrates of Cos (ca. 460 to 370 BC) and Galen of Pergamon (ca. 129 to 216) warned that for certain diseases (particularly plague) it was "dangerous to associate with those afflicted."

In AD 549, during an outbreak of plague in Constantinople, the Emperor Justinian enacted laws calling for the delay and isolation of travelers coming from regions where there was evidence of this disease. Therefore, in spite of the lack of appreciation that microbes could cause disease, people living in medieval Europe (and dying from the Black Death) recognized that the disease to which they were exposed was contagious and that the only way to preserve the public health would be to totally isolate the sick. When plague reached Europe in 1347, ports on the Adriatic and Mediterranean Sea were the first to deny entry to ships coming from pestilential areas, especially Turkey, the Middle East, and Africa. As early as 1348, Florence (the city where Boccaccio lived and wrote, and one of the most plague-ridden European cities) issued a restriction, called quarantine, on travelers and goods. The word "quarantine" comes from the Italian *quaranto giorni* and refers to the 40-day period during which ships entering the port were required to remain in isolation before the crew, cargo, and passengers were allowed to disembark. The Venetian Republic formally excluded "infected and suspected" ships in 1348, and in 1377 the first official quarantine was established in Dubrovnik. There persons and ships were isolated, usually on a nearby island, for 30 days to await signs of illness or continued good health. Later, other port cities established quarantine stations on shore or on neighboring islands. However, even with quarantine, and with 90% of the crew and passengers dying aboard the ship, the populations of port cities were decimated by plague. Today, thanks to the work of Yersin and Simond we understand the reason for this: the infected rats did not abide by the quarantine but left the ships via the docking lines! Quarantine was sometimes employed within the confines of the city itself, and the sick were often shut up in their homes with the uninfected members of the family and the flea-infested rats. This household quarantine did not prevent the spread of disease but, rather, resulted in a higher mortality.

Concerns Past

During the past two millennia, there have been three great bubonic plague pandemics. These resulted in social and economic upheavals unmatched by those due to armed conflicts or any other infectious disease. The first plague pandemic, called the plague of Justinian, came to the Mediterranean from an original focus in northeastern India or via Central Africa. The plague of Justinian was set off by a disruption of the pattern of plague transmission in its natural hosts—grass rats and gerbils. The disruption was caused by an unusual climatic effect—cooling and humid conditions—that precipitated a population explosion in these rodents. As a result, the larger numbers of rats and gerbils outstripped the bounds of their pastoral habitats and moved closer to human habitations. These rodents, which carried the efficient plague vector *Xenopsylla cheopis*, the Oriental rat flea, then acted as transmitters of plague. Once the grass rats and gerbils were crowded together with the domesticated black rats, the fleas jumped to the black rats and unleashed plague among human populations, who provided the transport system. From Lower Egypt, plague reached the harbor town of Pelusium in AD 540; from there it spread to Alexandria and then by ship to Constantinople, the capital of Justinian's empire. The historian Procopius of Caesarea described: "It embraced the entire world and blighted the lives of all men . . . great swellings . . . appeared in the groin and armpit . . . there was delirium, frantic restlessness and some became comatose. If the swellings were lanced there was the possibility of survival but most died in 5 days. It raged in the city of Constantinople [present-day Istanbul], the capital of the Roman Empire in the East, for months and so numerous were the corpses that they were cast into hollow towers of some incomplete fortifications." It is speculated that Justinian's General Belisarius was unable to accomplish Justinian's goal of reestablishing the Roman Empire in all its glory because of the outbreak of plague. This plague contributed to Justinian's failure to restore imperial unity due to a diminution of resources available, and hence it led to the failure of Roman and Persian forces to offer more than token resistance to the Moslem armies that swarmed out of Arabia in AD 634. Plague recurred every 3 or 4 years for decades, and its effects lasted well into the seventh century. It is estimated that by the year AD 600 the plague of Justinian had reduced the population by 100 million people—almost 50% of the population of Western Europe.

The historian William H. McNeill observed that the plague of Justinian resulted in "the perceptible shift away from the Mediterranean as the preeminent center of European civilization and the increase in the importance of more northerly lands." The Justinian plague and the subsequent outbreaks of plague over two more centuries marked the end of the Classical World—the Greek and Roman civilizations—and ushered in the Dark Ages. Plague so diminished trade in the Mediterranean that it left most countries with only a bartering economy; cities withered, feudalism grew, religion became more fatalistic, and Europe turned inward upon itself.

The source of the second pandemic, the Great Pestilence, about which Boccaccio wrote, was the "germs" from the Justinian plague that had moved eastward and remained endemic for seven centuries in marmots, large rodents native to the arid plateaus of central Asia (roughly corresponding today to Turkestan). When trappers came across these animals (dead or dying from plague), they were delighted to collect their fur. The pelts were sold to buyers from the West, and when the bales of marmot fur that had traveled along the caravan route known as the Silk Road were opened in Astrakhan and Saray, ravenous fleas jumped from the fur seeking a long-awaited blood meal. From Astrakhan and Saray, plague moved down the River Don to Kaffa, a seaport of southern Russia on the east coast of the Crimea. Kaffa was rat infested and provided a near-perfect breeding ground for the germ of plague. Many of the highly susceptible black rats in Kaffa also lived on the sailing ships. The holds of these ships were crawling with rats, and during the evening when the crew was asleep the rats took over the ship, running through the rigging. Infected rats rode the seas aboard these sailing ships—efficiently transporting plague to ports on the Mediterranean. Two of the most important ports connecting Kaffa with Europe were Constantinople (today Istanbul) and Messina (on the island of Sicily). Both cities became major foci for the further spread of plague. In Messina there was an outbreak of plague in 1347, and from there the Black Death crossed over to Tunis on the north coast of Africa and moved on to Spain by way of Sardinia by ship. By the time plague arrived in Spain, it had already arrived in the heart of Europe by sea—first in Genoa, then in Marseilles, and then in Pisa. Shortly thereafter it was in Great Britain, and it traveled to Scandinavia aboard a ship from London, carrying crew, wool cargo, and plague-infected rats, that docked in Bergen in 1349. By 1351 it was in Poland, and when it reached Russia in 1352 it had completed its circuit and a deadly noose had been secured around Europe.

Infected rats moved from port to port and country to country, spreading plague to the human populations living in the congested, filthy, rat-infested cities of Europe. Although the Black Death was undoubtedly the most dramatic outbreak of plague ever visited upon the western world, it did not disappear altogether. Between 1347 and 1722, plague epidemics occurred in Europe at infrequent intervals, occurring without introduction from caravans coming from Asia. In England, the epidemics occurred at 2- to 5-year intervals between 1361 and 1480. In the Great Plague of 1665, described in the diary of Samuel Pepys (and fictionalized in Daniel

Defoe's *Journal of the Plague Year*), at least 70,000 Londoners died (out of a population of 450,000).

The third plague pandemic began in the 1860s in China. Aided by swifter means of transportation—steamships and railways—it spread across the globe more rapidly than did the Black Death, and it is still with us. At present, most human cases of the plague are of the bubonic form, which results from the bite of a flea, usually the common rat flea, that has previously fed on an infected rodent. The bacteria spread to the lymph nodes (the armpits and neck but frequently the area of the groin), which drain the site of the bite, and these swollen and tender lymph nodes give the classic sign of bubonic plague, the bubo. Three days after the buboes appear, there is a high fever, the individual becomes delirious, and hemorrhages in the skin result in black splotches. Some contend that these dark spots on the skin gave the disease the name Black Death, whereas others believe that "black" is simply a mistranslation of "pestis atra," meaning, not black, but a "terrible" or "deadly" disease.

The buboes continue to enlarge, sometimes reaching the size of a hen's egg, and when they burst there is agonizing pain. Death can occur 2 to 4 days after the onset of symptoms. However, in some cases the bacteria enter the bloodstream. This second form of the disease, which may occur without the development of buboes, is called septicemic plague and is characterized by fever, chills, headache, malaise, and massive hemorrhaging, leading to death. Septicemic plague has a higher mortality rate than does bubonic plague. In still other instances, the bacteria move via the bloodstream to the lungs, leading to a suppurating pneumonia or pneumonic plague. Pneumonic plague, the only form of the disease that allows for human-to-human transmission, is characterized by watery and sometimes bloody sputum containing live bacteria. Coughing and spitting produce airborne droplets laden with the highly infectious bacteria, and others may become infected via inhalation. Pneumonic plague is the rapidly fatal form of the disease, and death can occur within 24 h of exposure. It is likely that this form of transmission (as well as transmission from human to human by the human flea *Pulex irritans*) was responsible for the rapidly spreading and devastating Black Death. Because pneumonic plague is highly communicable and exceedingly lethal, it frequently results in alarm and panic. In 1910 to 1911, pneumonic plague broke out in 60,000 people in Manchuria; in 1920 to 1921 there was a second large Manchurian pneumonic plague outbreak, affecting 10,000 people. Nearly all of those infected died. In 1994 there was an outbreak of pneumonic plague in Surat, India, that caused 1,400 deaths. Physicians fled the city, stating that "nothing could be done." Commercial airline flights and exports from India were cancelled, and it is estimated that the economy suffered a loss of $3 to $4 billion. In 2005, thousands sought refuge in neighboring countries when there was a suspected outbreak of pneumonic plague in the Congo.

Y. pestis is one of the most pathogenic bacteria: the lethal dose that kills 50% of exposed mice is a single bacterium injected intravenously. Typically, flea bites spread *Y. pestis* from rodent to rodent, but the bacterium can also survive for a few days in a decaying corpse and can persist for years in a frozen body. The reasons for the high degree of pathogenicity of *Y. pestis* remain unclear; however, some virulence factors have been identified. The genes of *Y. pestis*, located on one chromosome and three plasmids (circular DNA molecules), direct the synthesis of several proteins involved in disease production. One family of surface proteins, named Yops (Yersinia outer proteins), prevent the bacterium from being ingested by the host white blood cells (macrophages), thereby avoiding the immune mechanisms of the host; another, a toxin, kills other host cells; yet another protein is able to degrade fibrin, a material found in the blood clot, allowing the bacteria to move throughout the body by allowing them to escape from the clot at the site of the flea bite.

As with humans, the disease in fleas has a distinctive pattern. More than 80 different species of fleas are involved as plague vectors. Fleas are blood-sucking insects, and when a flea bites a plague-infected host (at the bacteremic/septicemic stage), it ingests the rod-shaped bacteria; these multiply in the blood clot in the proventriculus (foregut) of the flea. This bacterium-laden clot obstructs the flea's blood-sucking apparatus, and as a consequence the flea is unable to pump blood into the midgut, where it would be digested. As a result, the flea becomes hungrier, and in this ravenous state it bites the host repeatedly; with each bite, it regurgitates plague bacteria into the wound. In this way, infection is initiated. *Y. pestis* can also be pathogenic for the flea, and fleas with their foregut blocked rapidly starve to death. When the mammalian host dies, its body cools, and the fleas respond by moving off that host to seek another live warm-blooded host. However, if there is an extensive die-off of rodents, then the fleas move on to less preferred hosts such as humans, and so an epidemic may begin.

Y. pestis has been subdivided into three varieties—Antiqua, Medievalis, and Orientalis. Based on epidemiological and historical records, it has been hypothesized that Antiqua, presently resident in Africa, is descended from bacteria that caused the Justinian plague whereas Medievalis, resident in central Asia, is descended from bacteria that caused the Black Death; those of the third pandemic, and currently widespread, are all Orientalis. It is thought that *Y. pestis* probably evolved during the last 1,500 to 20,000 years because of changes in social and economic factors that were themselves the result of a dramatic increase in the size of the human population coincident with the development of agriculture. This increase in food supply for humans also became available for rodents, allowing rodent populations to expand as well. Increased numbers of rodents, coupled with changes in behavior, triggered the evolution of virulent *Y. pestis* from the enteric, food-borne, avirulent pathogen *Y. pseudotuberculosis*. This required several genetic changes: a gene whose product is involved in the storage of hemin resulted in blockage of the flea proventriculus and enhanced flea-mediated transmission, and other gene products (those encoding phospholipase D and plasminogen activator) facilitated blood dissemination in the mammalian body and allowed the infection of a variety of hosts by fleas.

Concerns Present

Plague is endemic in many countries in Africa, the former Soviet Union, and the Americas. In 2003 the World Health Organization (WHO) reported 2,118 cases and 182 deaths; two countries, Madagascar and Democratic Republic of the Congo, accounted for 2,025 of these cases and 177 deaths. According to the WHO, the actual number of cases in the world is probably much higher than reported due to a reluctance on the part of many countries to declare that they have had an outbreak of "the Black Death," as well as to poor diagnosis because the clinical picture was not clear and there was no laboratory confirmation. In December 2004, pneumonic plague broke out among workers in a diamond mine in the Democratic Republic of the Congo; by mid-March, when the outbreak was brought under control, 130 people had been infected and 57 were dead. In 2005, two Tibetans died and three others recovered from plague after eating marmot meat.

Worldwide, on average in the last 50 years, ca. 2,000 cases of plague have been reported annually. Outbreaks of plague in developed countries are rare (e.g., 12 cases per year in the United States). Most human cases in the United States occur in northern New Mexico, northern Arizona, southern Colorado, California, southern Oregon, and far-western Nevada. The last urban plague epidemic in the United States occurred in California in 1924 to 1925.

For most developed countries the greatest threat from plague may come from the intentional release of *Y. pestis* for nefarious purposes. The earliest example of the use of plague as a biological warfare weapon was in 1346 during the siege of the Baltic Sea city of Kaffa by the Mongols. According to an account by Gabriel de Mussis, who was an eyewitness, the siege of Kaffa was stalled for several years. When plague broke out among the Mongol army, their leader ordered that catapults be used to hurl the plague-ridden dead bodies over Kaffa's walls. It was reasoned that the diseased "flying corpses" would infect the Genoese traders who had taken refuge in Kaffa. The besieged Genoese threw the bodies over Kaffa's ramparts and into the sea, but plague had already infected the inhabitants of Kaffa. When the Mongols abandoned their siege, the Genoese traders were free to leave; in 1347, when they set sail for Italy, they were accompanied by plague-infected rats. Their first docking was at Messina, which became the portal through which plague entered the port cities along the Mediterranean, eventually leading to the massive die-off of the Black Death.

During the 1930s the Japanese military attempted to spread plague in China by dropping plague-infected fleas from aircraft. Human infections by flea bite usually result in the bubonic form, and most often this transmission is a dead end with no further human-to-human infection; however, if the bacteria enter the lungs (and recent studies show this to occur in 12% of cases), these persons may acquire secondary pneumonic plague, which is transmissible to others by coughing, spitting, and even breathing and can lead to primary pneumonic plague. Because primary pneumonic plague is transmissible from person to person and infections can be maintained in the absence of rodents or fleas, it was developed in the 1990s by the Soviets into an aerosol that could be used as a biological warfare agent. It is unknown what happened to the stockpiles of this material.

Let us for a moment imagine that some of that aerosolized plague fell into the hands of a terrorist organization and was covertly released in City Stadium, where a few thousand fans were attending a championship baseball game. Can we imagine the consequences? Just such an exercise for top public health and governmental officials was played out in the fictional scenario "Plague on Your City." A day after the release of the aerosol, the City Health Department would begin to receive information that an increasing number of people have been appearing at local hospitals with coughs and fever. Initially it is regarded as an outbreak of the summer flu. Later that day there are more than 500 people so affected, and 24 h later 25 people have died. Specimens are taken from the sick, and samples are sent to the Centers for Disease Control and Prevention (CDC) in Atlanta, Ga. After the CDC and the City Health Department determine that the infectious agent is not the flu virus but the bacterium of plague, a CDC team is dispatched to the city to coordinate a response. First, a public health emergency is declared. Hospitals that only a day ago believed they were dealing with influenza now begin recalling staff and instituting emergency plans and treatment protocols. By late afternoon of day 2, hospital staff begin to call in sick and antibiotics and ventilators become scarce. Second, the Governor issues an executive order that the entire city be quarantined. Antibiotics are also commandeered. Third, the public is informed that there has been a terrorist attack with an outbreak of plague and that they should seek treatment if they fall ill or have been in contact with a plague case. Panic and rioting begin to be reported as people storm health facilities to demand antibiotics. Fourth, people who are healthy are told to stay indoors, to avoid public gatherings, and to use dust masks to prevent the spread of disease. Despite these rapid response measures, by the beginning of day 3 there are 1,800 cases. Most of these are in the United States, but cases are reported from London and Tokyo. Of the infected persons, 400 have died. During day 3, hospitals cannot cope with the influx of sick patients. Medical care begins to shut down. There are difficulties in getting antibiotics to the facilities in need. A distribution plan is yet to be instituted. To limit the spread of disease, the state border becomes a cordon sanitaire. People begin to worry how food and supplies will get to the city. Despite the prohibition of travel into and out of the state, there are reports of over 3,000 patients with pneumonic plague, of whom 800 have died. By day 4 the number of cases has soared to 4,000, with 1,000 deaths. At this point the exercise was terminated and critiqued.

The critique of this exercise revealed that with an outbreak of plague, there must be decisive leadership, priorities for the distribution of scarce resources must be established, and adequate health care facilities must be developed along with the formulation of sound principles for disease containment. The lessons from this mock-attack scenario are clear: rapid and effective planning and preparedness at all levels of government and by public health agencies are necessary to prevent a recurrence of the Black Death.

Consequences

The mortality and morbidity from plague have been significantly reduced in the 21st century. However, the disease has not been eradicated; it remains endemic in regions of Africa, Asia, and North and South America. From 1983 to 1997, 28,570 cases with 2,331 deaths in 24 countries were reported to the WHO. In 1997 the total number of cases reported by 14 countries to the WHO was 5,419, of which 274 were fatal. Epidemics occurred in Madagascar in 1991 and 1997; in Malawi, Zimbabwe, and India in 1994; and in Zambia and China in 1996. In contrast, there were four cases and one death in the United States in 1997.

Small mammals such as urban and sylvatic (or wood) rats as well as squirrels, prairie dogs, rabbits, voles, coyotes, and domestic cats are the principal hosts for *Y. pestis*. The increase in plague in Vietnam during the Vietnam War has been attributed to deforestation that resulted in the movement of these sylvatic hosts into areas where humans lived and fought. In the United States, plague has been encountered when houses have encroached on rural areas that harbor plague-infected rodents (sylvatic plague) such as prairie dogs, ground squirrels, mice, and rats. Infection may also be acquired from domestic pets (cats and dogs) that have come in contact with prairie dogs, and the pets can transmit disease to humans via fleas or in the pneumonic form. In the United States there has been an increase in the number of human cases of pneumonic plague in persons, especially veterinarians, who have been exposed to infected cats.

Plague can be controlled by surveying wild populations for infections, monitoring die-offs in the rodent population, making plague-infested areas known to the public, reducing the appeal of residential areas to rodents, and treating rodent burrows with carbaryl dust or bait stations to kill fleas. In some cases rodenticides have been used. Because of sylvatic plague and enzootic infections (i.e., an outbreak of a disease in animals other than humans), complete eradication of plague is not possible.

Although human disease is rare, a feverish patient who has been exposed to rodents or fleabites in plague-endemic areas should be considered a possible plague victim. The potential for the spread of pneumonic plague has been increased by air travel. Passengers who have fever, cough, or chills and come from areas where plague is endemic should be placed in isolation and, if necessary, treated. The diagnosis of plague has remained virtually unchanged since the days of Yersin: Gram staining and culture of bubo aspirates or sputum. The bacteria can also be grown in the laboratory on blood and MacConkey agar.

Unless specific treatment is given, the condition of a plague-infected individual deteriorates rapidly and death can occur in 3 to 5 days. Untreated plague has a mortality rate of more than 50%. Streptomycin, gentamicin, tetracyclines, and chloramphenicol can be used to treat pneumonic plague. Antibiotic resistance, though rare, has been reported. Travelers at high risk in epidemic areas should be advised to consider prophylactic measures, i.e., tetracycline or doxycycline during periods of exposure. Insect repellents are also recommended.

Natural plague poses no threat to most of the developed world; however, a deliberate plague attack would be a different story. An outbreak of plague, whether by human or natural causes, cannot enter the human population without there being intimate involvement of people in its transmission: by human contact with rats and their infected fleas, by fleas jumping from human to human, or by traveling while infected, thus contaminating the air with plague-laden droplets. Indeed, the power of plague lies in its ability to spread; only if the chain of transmission is interrupted can an epidemic be halted. Since the time of Boccaccio, we have learned that the most effective public health measure to contain a plague pandemic is rapid-response quarantine. Today we would supplement quarantine with careful monitoring, the use of face masks, and the distribution of antibiotics. "In the time of a new plague people will have to remember despite their terror that no epidemic disease is more susceptible to quarantine. The very notion of quarantine has an ancient horror; in the brutal lockups of the Renaissance, people starved to death, or died in miserable isolation. But today, people will need to impose upon themselves the requirement, in the event of an outbreak, to stay out of public places and, if at all possible, to stay in their homes. We would need gauze masks to run errands; we would need access to food, water, and medical care. None of this would be easy to arrange. But terror and flight have always been plague's handmaids. We would need to resist the age-old desire to flee, to head to safer country, somewhere else in the hills, in the woods, in some other city. We would need to remember that plague . . . cannot linger in the environment. A plague attack would be over before the very first infections ever appeared. Therefore, flight makes no sense. Self-imposed isolation does."

6

Syphilis: the Great Pox

It was 1748, and London was the center of the universe . . . they talked of . . . Voltaire, Blackstone was a judge, Chippendale a furniture maker. Samuel Johnson wrote . . . of the permanent and certain characteristics of the mind. The pox was sexually transmitted. The pox, if you were unlucky, could rot the organ of manhood. Some said the pox was one disease others said it was really two similar diseases. In John Hunter's mind . . . it wasn't transmitted by miasmas at all. The pox must be caused by . . . a putrid liquid of some sort. John Hunter had the pox all figured out. In his opinion there were two forms. If it began as a pimple on the penis, it took one specific course. The other form was seen when . . . the interior of the urethra was affected . . . and produced a liquid discharge. A patient with the pox stood before him . . . a tiny drop of yellowish liquid hung from the tip of his penis . . . it was a classic case . . . of the pox. To prove that the pox was a single disease all he needed to show was that the pus from this patient, a "wet" case, could produce a chancre, or a "dry" case in another penis. John only had one healthy penis handy, so he used it. While the patient stood before him, he took the droplet on a lancet and transferred it to the glans of his own penis. Then, through the droplet, he stabbed himself with the lancet . . . and again. He squeezed the small cuts open, so that the liquid . . . of the pox could take hold. He gave the man a bit of mercury . . . to rub . . . into his thighs. That Sunday he confided in his notebook that his penis began tingling . . . there was also a redness where the lancet had pierced the skin. By the following Tuesday there were two pimplelike chancres . . . A runny disease had produced a dry disease, proving, once and for all . . . the pox was but one disease. He rubbed mercury into his thighs and the chancres disappeared. Three months later he got a skin rash and . . . he rubbed a massive dose of mercury in his thighs and the symptoms, he wrote in his notebook, disappeared for good. The inevitable happened in 1793 . . . he died. John Hunter had been wrong. John Hunter had died of . . . syphilis—and by his own hand.

During the time of John Hunter (1728 to 1793), the origins of the Great Pox were obscure. One of the most popular theories was that some of the 44-member crew of Christopher Columbus (1451 to 1506) contracted the disease from Native Americans during their visit to the Americas. On their arrival in Spain in 1493, some of these men joined the army of Charles VIII of France, who launched an invasion of Italy in 1494 and besieged the city of Naples in 1495. During the siege his troops fell ill with the Great Pox, and this forced their withdrawal. With the disbanding and dispersal of the soldiers of Charles VIII, who themselves had been infected by the Neapolitan women, the disease spread rapidly through Europe. In the spring of 1496 some of these mercenaries joined Perkin Warbeck in Scotland and, with the support of James IV, invaded England. The pox was evident in the invading troops. Within 5 years of its arrival in Europe, the disease was epidemic: it was in Hungary and Russia by 1497 and in Africa and the Middle East a year later. The Portuguese, the earliest to actually receive the infection, carried it around the Cape of Good Hope with the voyages of Vasco de Gama to India in 1498. It was in China by 1505, in Australia by 1515, and in Japan by 1569. Eventually the Great Pox was carried to every continent save Antarctica.

The French called this pox "the disease of Naples," blaming the Italians for it; the Italians, however, called it "the French disease." The basis for these various names is that the undisciplined troops of Charles VIII, during their retreat from Italy, carried the disease back to their homelands in many parts of Europe; shortly thereafter, it came to be called after the national origins of people who were disliked and considered unclean—the Russians called it "the Polish disease," the Japanese called it "the Chinese disease," and the English called it "the Spanish disease."

Victims suffered with fevers, open sores, disfiguring scars, disabling pains in the joints, and gruesome deaths, leading Joseph Grunbeck (1473 to 1532) of Germany to write in the late 15th century, "In recent times I have seen scourges, horrible sicknesses and many infirmities affect mankind from all corners of the earth . . . a disease which is so cruel, so distressing, so appalling that until now nothing so horrifying, nothing more terrible or disgusting, has ever been known on this earth."

Because the specific agent responsible for the Great Pox would not be known for 5 centuries, there was ample time to speculate on its origins. Some proffered an astrological basis: in 1484 Mars, Jupiter, and Saturn were in conjunction with Scorpio, the constellation most commonly associated with sexuality. Others believed that it was God's punishment for human sexual excesses. As a consequence, public bathhouses were closed, there was a reduction in the level of prostitution, and distrust arose between friends and lovers. It has been suggested that the fashion of wearing wigs and gloves is also related to the Great Pox since these would conceal the disease-induced sores. In 1530 Girolamo Fracastoro (1483 to 1553) named the Great Pox "syphilis" after a fictitious shepherd, Syphilis, who, by cursing Apollo, so angered the god that he afflicted the shepherd with a new disease: Fracastoro also claimed that syphilis was an infectious disease caused by "seeds of contagion." The idea that syphilis was the result of seeds of contagion was derided by Fracastoro's contemporaries because the intellectual climate of the time would not support such a notion and the seeds could not be detected by the senses of smell, sight, or touch. However, 300 years later, Fracastoro's seeds of contagion were found.

In 1905 Fritz Schaudinn (1871 to 1906) and Erich Hoffmann (1868 to 1959) in Germany, using their powerful light microscopes, discovered a bacterium in the fluid expressed from syphilitic chancres. Because of its twisted shape Schaudinn and Hoffmann named it *Treponema* ("trep," meaning corkscrew, and "nema," meaning thread in Latin), and since it stained so poorly they named its kind *pallidum* (from "pallid," meaning pale). In 1913, when Hideyo Noguchi (1876 to 1928) isolated the same microbe from the brains of patients with insanity and paresis due to late-stage syphilis, it became clear that all of the stages of the disease were linked to one "seed of contagion," *T. pallidum.*

The treponemes that cause the human diseases yaws, pinta, treponarid (also called bejel), and venereal syphilis appear identical under the microscope, and until 1998 they could not be distinguished except by their clinical manifestations. Pinta, restricted to the skin, changes the skin's pigmentation, and the treponemes are spread from person to person by being introduced into breaks in the skin. Yaws exists in warm, moist climates, especially in places characterized by poor hygiene, and is transmitted primarily in children by skin contact; it is not a benign disease and can spread throughout the body via the blood to cause disfigurement of the face and bones. Treponarid, transmitted via objects contaminated with saliva—drinking vessels and kitchen utensils—as well as by mouth-to-mouth contact, can be destructive to tissues such as the skin and bones later in life.

In 1998 the complete genome of *T. pallidum* was decoded. Its genome is quite primitive, contains only 1,000 genes (compared to the human genome, which contains ~25,000), and has a limited capacity for the synthesis of critical metabolites. As a consequence, sugars, amino acids, and nucleic acid precursors have to be acquired from the host through 18 special transporters. Potential virulence factors include genes that encode proteins for punching holes in the host cell, as well as sticky substances that allow the treponeme to attach to the mucous membranes. Duplicated genes allow it to change its surface proteins, enabling the treponeme to evade the immune response. Although the closely related treponemes that cause yaws *(Treponema pallidum* subsp. *pertenue),* treponarid (*Treponema pallidum* subsp. *endemicum*), and pinta (*Treponema carateum)* are >95% genetically alike, a single gene (called *tpp15*) marks *T. pallidum* as distinct. This is the genetic signature of venereal syphilis. *T. pallidum* and its relatives are extremely sensitive to heat because they lack heat shock genes. Lacking the enzymes that protect against oxygen toxicity, they can survive only in places where the oxygen levels are low. Also, their exquisite sensitivity to dehydration forces them to live in moist places.

Origins

The origin of venereal syphilis has been a source of controversy for centuries. A popular theory, called the Columbian theory, claims that the disease was a New World import brought to Europe by Columbus' sailors. Supporting this theory was the outbreak of the disease in Europe at the time of the return of Columbus and his crew from the New World and the suggestion (which was never actually confirmed) that Columbus' sailors had syphilitic lesions. The strongest evidence for the Columbian theory was provided in the bones of skeletons: bone lesions characteristic of syphilis—scrimshaw patterns and saber thickenings on the lower limbs—were found in Amerindian skeletons older than AD 1500, yet for many years skeletons recovered in Europe and China, and dated to be earlier than AD 1500, showed no such lesions. However, recent finds and interpretations cast doubts about the Columbian theory. There are several older written reports stating that Columbus' crew and the Amerindians were healthy, and the bone lesions in the Amerindian skeletons have now been found not to be specific for venereal syphilis but more likely represent the presence of the related (but nonvenereal) disease, yaws. Indeed, all pre-Columbian Amerindian skeletons lack the moth-eaten-appearing skull lesions (caries sicca) generally accepted as diagnostic for late-stage syphilis. In the past few decades Old World remains (dated to be older than 1495) have been found, and these bear the caries sicca signature of venereal syphilis. In 2000 it was reported that 8 of 245 skeletons unearthed from a medieval monastery, known as Blackfriars, in Hull, England, show the telltale signs of caries sicca. Carbon dating of the bones established the date of death to be between 1300 and 1420—at least 70 years before Columbus' voyage. In 1994, Henneberg and Henneberg reported finding caries sicca in the remains of 47 individuals found in the Greek colony of Metaponto, Italy, dated to 600 BC. At a French site, Lisieux, a skeleton dated from the fourth century has caries sicca and saber thickenings of the long bones, and similar finds have been made at Rivenhall, at a friary graveyard in Gloucester, and in another four skeletons at Norwich, all in England, dated to be mid-15th century or earlier. A skull from the Song dynasty (AD 960 to 1279) in Fujiang, China, has lesions similar to caries sicca. This evidence strongly suggests that a syphilis-like disease was present in Europe and Asia for thousands of years and allows for another theory for the origins of the Great Pox.

Humans probably first acquired their treponeme infections from animals living in tropical Africa. Initially, the human disease resembled that found in present-day baboons and was spread from person to person by skin contact. A million years ago these pinta-like treponemes mutated into a form similar to that which causes yaws. Indeed, a skeleton of African *Homo erectus,* dated to be more than 1.6 million years old, shows the telltale scars of yaws. Initially yaws was restricted to the tropical areas of Africa and the associated landmasses where the climate is moist and warm; it was transmitted primarily in children by skin contact. When the populations of *H. erectus* migrated out of Africa and passed across the Bering Straits, they carried yaws with them. It became isolated in the tropics of the Americas, places where the humidity is high and hygiene is poor. (Much later it would be reintroduced into the Americas from Africa through the slave trade.) As human populations began to penetrate into temperate and drier regions, the presence of a cooler and drier climate and the wearing of clothing provoked (or so it is theorized) the treponemes to mutate again, allowing them to now invade the mouth and throat and to cause skin lesions similar to treponarid. Transmission occurred by objects contaminated with saliva—drinking vessels and kitchen utensils—as well as by mouth-to-mouth contact. In treponarid and yaws, the primary site of infection may have resembled impetigo (a skin disease caused by streptococci) or herpes simplex (cold sores), but in later life there could be destruction of tissues such as the skin and bones.

Before the beginning of the 16th century there was a low level of treponemal infections—similar to yaws and treponarid—that spread far and wide across the globe and was transmitted mostly by nonvenereal means. Childhood infections with these treponemes may have resulted in partial immunity to other treponeme infections. Indeed, had a mutation arisen that resulted in a venereally transmitted treponeme, it would have been at an evolutionary disadvantage in a sexually active population since that population would be resistant to such an infection. However, even when such a rare venereal variant did occur, it would have been unable to spread very far. The transition from a nonvenereal to a venereal transmitted infection, with the possibility for widespread dissemination, required a dramatic change in the lifestyle of the human population. Such a change took place in the 15th and 16th centuries when environmental conditions in Europe became sufficiently favorable for the previously disadvantaged mutants of venereally transmitted treponemes to prosper. At the royal courts there was sexual liberty; public bathhouses and other unhygienic establishments were common in the filthy and crowded urban areas; and prostitution was on the increase. Leprosaria, institutions where syphilitics might have also been interred, were closed by papal order, and the sick were allowed to move to other communities. With the fall of Islamic Granada in 1492 and the expulsion of the Muslim and Jewish populations, increased contact with other peoples took place. There was war and extensive trade: armies and companies of mercenaries roamed the battlefields of love and war, facilitating the spread of the venereally transmitted mutant. Under these conditions, Columbus' crew came to be the vector for venereal syphilis, not its source. The effect was that venereal syphilis broke out in Europe in the late Middle Ages and the beginning of the Renaissance.

Syphilis is a disease of cities, and with the high population density, changes in social habits, wearing of clothing, less frequent sharing of eating utensils, and unique opportunities for rapid transmission, the propagation of the milder forms of treponemes was reduced, allowing transmission of only the more virulent venereal forms. The accumulating evidence now favors the notion that syphilis probably did exist in Europe and Asia in a nonvenereal form prior to 1493 but then became a virulent and rapidly spreading disease. Favoring this idea was the severity of the outbreak—indicative of a new import. Indeed, from 1494 to 1516 the first signs were described as genital ulcers followed by a rash, and then the disease spread throughout the body, affecting the gums, palate, uvula, jaw, and tonsils, eventually destroying these organs. The victims suffered pains in the muscles, and there was early death, indications of an acute disease. From 1516 to 1526 two new symptoms were added to the list: bone inflammation and hard pustules. Between 1526 and 1560 the severity of symptoms diminished and thereafter the deadliness continued to decline, but from 1560 to 1610 there was another sign: ringing in the ears. By the 1600s, "the Great Pox" was an extremely dangerous infection, but those who were afflicted did not suffer the explosive attacks that had been seen in the 1500s. Either because people were developing increased resistance or because the bacterium was showing reduced virulence, the disease became an easily transmissible, mild, nonfatal, nondescript acute illness that progressed into a debilitating chronic illness.

Clinical Signs

There is enormous variation in the disease symptoms, and so syphilis has been called "the Great Imitator." Infections are characterized by three stages. The chancre stage is the earliest clinical sign of disease. After initial contact, *T. pallidum* rapidly penetrates intact mucous membranes or microscopic breaks in the skin; within hours, it enters the lymph and blood vessels to cause a systemic infection long before any other signs appear. In ~21 days (range, 3 to 90 days), a painless, pea-sized ulcer, called a chancre, appears at the site where the treponemes entered the body. The chancre, a local tissue reaction, can occur on the lips, fingers, or genitals. Multiple lesions occur in 30% of cases. If untreated, the chancre usually disappears within 4 to 8 weeks, leaving a small, inconspicuous scar. The individual frequently does not notice either the chancre or the scar, although there may be lymphadenopathy. (Because the lesion—the chancre—is larger than that of smallpox, syphilis has been called "the Great Pox.") At this stage the infection can be spread by kissing or touching a person with active sores on the lips, genitalia, or breasts and through ingestion of breast milk.

The secondary (or disseminated) stage usually develops 2 to 12 weeks (mean, 6 weeks) after the chancre; however, this stage may be delayed for more than a year. Treponemes are present in all the tissues but especially the blood, and there is a high level of syphilis antigen. Serologic tests (such as the Wasserman, VDRL, and RPR [rapid plasma reagin] tests) are positive. There is now a general tissue reaction: headache, sore throat, a mild fever, and, in 90% of cases, a skin rash. The skin rash might have been mistaken for measles, smallpox, chicken pox, or some other skin disease. In 40% of cases, the central nervous system is involved. The highly infectious secondary stage does not last very long. It is followed by the hidden or early latent stage, in which the individual appears to be symptom free;

i.e., there are no clinical signs. Indeed, the most dangerous time is during the early latent stage, when the syphilitic individual is able to transmit the disease to others while appearing to be disease free. Transmission via by blood transfusion is rare because *T. pallidum* is quite fragile and does not survive outside the body longer than 24 to 48 h even under the most favorable conditions, i.e., in blood.

The infection may continue to progress, and after about 2 years the late latent (tertiary) stage develops. It should be noted, however, that 70% of people with untreated syphilis never develop clinically evident late-stage syphilis. Although treponemes are still present in the body during tertiary syphilis, the individual is no longer infectious through sexual contact 4 years after initial contact. Complications develop in 15 to 40% of untreated patients with tertiary syphilis: destructive ulcers (gummas) appear in the skin, muscles, liver, lungs, and eyes; the heart is damaged; and the aorta is inflamed—in severe cases, the aorta may rupture, causing death, as happened to John Hunter. The spinal cord and brain can become involved, causing incomplete paralysis (paresis), complete paralysis, and/or insanity, accompanied by headaches, pains in the joints, impotence, and epileptic seizures. Most untreated patients die within 5 years after showing the first signs of paralysis and insanity.

The course of a syphilitic infection is well documented. In 1890 to 1910, 1,404 untreated patients were studied in the Oslo study. However, only 24% of the patients were autopsied. Of these, 28% showed signs of tertiary-stage syphilis. Symptomatic neurosyphilis developed in 10% of males and 5% of females; 13% had cardiovascular disease, and syphilis was the primary cause of death in 17% of men and 8% of women. P. D. Rosahn studied the cadavers at the Yale School of Medicine from 1917 to 1941. Of the 4,000 autopsied cadavers, ~10% showed evidence of syphilis. Of the 77 patients with untreated syphilis, 83% had cardiovascular disease, 20% were presumed to have died from complications related to the disease, and 9% had neurological complications.

Between 1932 and 1972, the now infamous Tuskegee Syphilis Study was designed to track untreated syphilis in Macon County, Ala., where the overall infection rate was 36%. The U.S. Public Health Service began the study and enlisted 399 poor, black sharecroppers living in Macon County, all with latent syphilis. Cooperation was obtained by offering financial incentives such as free burial service, on the condition that they agreed to an autopsy; the men were also given free physical examinations, and a local county health nurse, Eunice Rivers, provided them with incidental medications such as "spring tonics" and aspirin whenever needed. The men (and their families) were not told they had syphilis; instead, they were told they had "bad blood," and annually a government doctor would take their blood pressure, listen to their hearts, obtain a blood sample, and advise them on their diet so that they could be helped with their bad blood. However, these men were not told that they would be deprived of treatment for their syphilis and they were never provided with enough information to make an informed decision. The men enrolled in this study were denied access to treatment for syphilis even after penicillin came into use (1947); instead, they were left to degenerate under the ravages of tertiary syphilis. By the time the study was made public, largely through James Jones' book *Bad Blood* and the play "Miss Evers' Boys," 28 men had died of syphilis, 100 others were dead of related complications, at least 40 wives had been infected, and 19 children had contracted the disease at birth.

The Tuskeegee Study found nothing new: there were signs of cardiovascular disease in 46% of the syphilitics and 24% had an inflammation of the aorta, whereas the percentages in the controls were 24% and 5%, respectively. Bone disease was found in 13% of the syphilitics but in only 5% of the controls. The greatest differences were seen in the central nervous system; 8% of the syphilitics had signs of disease compared to 2% of the controls. After 12 years the death rate in the syphilitics was 25% versus 14%, at 20 years it was 39% versus 26%, and at 30 years it was 59% versus 45%. The Tuskegee Study has come to symbolize racism in medicine, ethical misconduct in medical research, paternalism by physicians, and government abuse of society's most vulnerable: the poor and uneducated. On 16 May 1997 the surviving participants in the study were invited to a White House ceremony and President Bill Clinton said, "The United States government did something that was wrong—deeply, profoundly, morally wrong. It was an outrage to our commitment to integrity and equality for all our citizens. Today all we can do is apologize but . . . only you have the power to forgive."

"Catching" the Great Pox

At the primary (chancre) stage, the disease can be spread by kissing or touching a person with active sores. During the secondary stage, which usually does not last very long, the skin lesions render the individual infectious. In the early latent stage there are no clinical signs; however, the individual remains infectious. Syphilis can be transmitted from the mother to the developing fetus via the placental blood supply, resulting in congenital syphilis; this is most likely to occur when the mother is in an active stage of infection. Fetal death or miscarriage usually does not occur until after the fourth month of pregnancy at the earliest. Repeated miscarriages after the fourth month are strongly suggestive of but not unequivocal proof for syphilis. A diseased surviving child may go through the same symptoms as the adult, or there may be deformities, deafness, and blindness. Of particular significance is that some offspring who are congenitally infected may have Hutchinson's triad: deafness, impaired vision, and a telltale groove across peg-shaped teeth, first described in 1861 by the London physician Jonathan Hutchinson. (If the mother is treated during the first 4 months of pregnancy, the fetus will not become infected.)

Chemotherapy

Treatments for the Great Pox varied with the times. George Sommariva of Verona, Italy, tried mercury for the treatment of "the French pox." By 1497, mercury was applied topically to the suppurating sores or was taken in the form of a drink. The treatment came to be called "salivation" because the near-poisoning with mercury salts tended to produce copious amounts of saliva. Another treatment involved guaiacum (holy wood) resin from trees (*Gauiacum officianale* and *G. sanctum*) indigenous to the West Indies and South America. Although the resin was a useless remedy, it was probably popularized to lend further credence to the American (Columbian) origins of the pox. However, few adherents of the Columbian theory paid attention to the fact that guaiacum was introduced as a treatment in 1508, fully 10 years before the first mention of the West Indian origin of syphilis.

By the end of the 19th century it was estimated that 10% of the population of Europe had syphilis, and by the early 20th century it was estimated that one-third of all patients in mental institutions could trace their neurological symptoms to the disease. In 1909 the chemist-immunologist Paul Ehrlich developed a drug that was effective in reducing the severity of syphilis. How did this happen, and how did that discovery provide the foundation for the discovery of antibiotics such as penicillin to cure the Great Pox? It all began with the German dye industry. By the early 1800s, the large-scale commercial manufacture of illuminating gas (ethylene) for lighting purposes provided the raw materials necessary for the synthesis of dyes. In England, illuminating gas was obtained by distillation from coal; however, it was not the gas, but the tar, that was the real goal. The British navy, as a result of the loss of their American colonies that had supplied it, needed tar. Unwilling to be an economic captive to the colonies, the British attempted to become self-sufficient and to make the tar from coal. Coal gas was the first by-product of coal distillation and initially was of little value. However, after 1812 the value of the gas outweighed that of tar. As a consequence of the stepped-up production of illuminating gas, the tar would have accumulated in great quantities had not the chemists discovered a new use for it. By boiling and distilling the tar, it was possible to obtain light and heavier oils, called creosote and pitch, respectively. The creosote was used for preserving wood, and the pitch was sold for asphalt manufacture. A sample of the light oil was sent to Germany for analysis by a brilliant young German chemist, August Hoffmann, who found that it contained benzene and aniline. Hoffmann made a variety of dyes by starting with aniline, and by 1850, due to his efforts and those of his students, Germany became the center of the European dye industry, producing aniline dyes in all shades of the rainbow. It was therefore natural that another German, Paul Ehrlich (1854 to 1915), would become interested in dyes and their uses.

Ehrlich studied at the University of Strasbourg, where his tutor, Professor Waldeyer, indulged his experimentation with various kinds of dyes, and "little Ehrlich," as Waldeyer called him, discovered a new type of cell—the mast cell. (Later, it would be found that mast cells secrete the chemical histamine, making nearby capillaries leaky and in some cases provoking an allergic response. Drugs designed to blunt this effect are called antihistamines.) During this time he also found that certain dyes (particularly methylene blue and neutral red) were taken up by living cells. Ehrlich graduated as a doctor of medicine at age 24. His thesis was on "The value and significance of staining tissues with aniline dyes." Because certain dyes stained only certain tissues and not others, it suggested to Ehrlich that the binding showed chemical specificity. This notion became the major theme in his scientific life and led to a search for "magic bullets"—drugs—that would specifically bind to and kill parasites.

In 1883, working with Emil von Behring in Robert Koch's laboratory in Berlin, Ehrlich investigated the relationship of neutralizing antibodies to toxins and was able to standardize diphtheria antitoxin so that it could be used in the treatment of human infections. As a result of this work, the German government established the Royal Institute for Experimental

Therapy in Frankfurt to provide such antitoxins and vaccines; Ehrlich was appointed Director in 1899. His work on immunity suggested that only when there was a specific interaction between an antibody and an antigen—as a specific key fits into and opens a particular lock—did neutralization occur. Visits to the Hoechst factory near Frankfurt brought Ehrlich face to face with German dye works, where a profusion of synthetic analgesics, antipyretics, and anesthetics were being made. It seemed logical to him that since such substances were effective by acting on specific tissues (in much the way dyes did), it should be possible to synthesize other small molecules that would act differentially on tissues and parasites. He recognized that immunotherapy—the use of vaccines—was a matter of strengthening the defense mechanisms of the body whereas with drug therapy there would be a direct attack on the parasite. He also observed that immunotherapy involved an animal making the large, unstable protein molecule (antibody) whereas drug therapy involved making a small and stable molecule in the laboratory.

Ehrlich wrote, "Curative substances—a priori—must directly destroy the microbes provoking the disease; not by an 'action from distance,' but only when the chemical compound is fixed by the parasites. The parasites can only be killed if the chemical has a specific affinity for them and binds to them. This is a very difficult task because it is necessary to find chemical compounds, which have a strong destructive effect upon the parasites, but which do not at all, or only to a minimum extent, attack or damage the organs of the body. There must be a planned chemical synthesis: proceeding from a chemical substance with a recognizable activity, making derivatives from it, and then trying each one of these to discover the degree of its activity and effectiveness. This we call chemotherapy." Ehrlich's first chemotherapeutic experiments were carried out in 1904 with mice infected with the trypanosomes that cause African sleeping sickness. He was able to cure these mice of their infection by injections of a red dye he called trypan red. This was the first man-made chemotherapeutic agent. At first, trypan red aroused some interest; however, because the drug was inactive in human sleeping sickness, Ehrlich turned his attention to organic arsenicals. He began with the compound atoxyl—an arsenic-containing compound that was supposedly a curative for sleeping sickness—but he found that it was useless: it destroyed the optic nerve, so that when patients were treated with the drug, not only were they not cured of sleeping sickness, they became blind. In 1906 he prepared compound 418, arsenophenylglycine, which killed trypanosomes, and then he prepared the

606th derivative, a compound called 606, a dioxydiaminoarsenobenzene (= arsphenamine), which also killed trypanosomes in the animal but did not do so in the test tube. This finding, claimed Ehrlich's critics, indicated that this drug had no direct effect on the parasite but, rather, that its action was due to a stimulation of the host's natural defenses.

Most of Ehrlich's contemporaries ridiculed him, calling him "Dr. Phantasmus"; furthermore, they were unimpressed with his dye-based research and thought that nothing of value would come from it. Indeed, for 5 years Ehrlich was unable to produce a single drug that was of use in humans. But in 1905, all that changed. That was the year when Schaudinn and Hoffmann described the germ of syphilis, and shortly thereafter Sahachiro Hata, working at the Kitasato Institute for Infectious Disease in Tokyo, discovered how to reproduce the disease in rabbits. It was intuition rather than logic that led Ehrlich to believe that arsenicals would kill *Treponema*. (His reasoning was that since trypanosomes and treponemes are both active swimmers, they must have a very high rate of metabolism, and an arsenical would kill the parasite by crippling its energy-generating ability.) And now, thanks to Hata's rabbit model for human syphilis, Ehrlich's drugs could be tested for their chemotherapeutic effectiveness. In the spring of 1909, Professor Kitasato of Tokyo sent his pupil, Dr. Hata, to Germany to work with Ehrlich, and a year later Hata successfully treated syphilis in rabbits with compound 606. The synthesis of compound 606 was reported in 1912; the molecule was patented, manufactured by Hoechst, and sold by them under the name "salvarsan." The announcement of a cure for syphilis in humans was taken up by the newspapers, and overnight Ehrlich became a world celebrity. It was dangerous to make salvarsan because the ether vapors used in its preparation could cause fires and explosions, and it was unstable—a trace of air changed it from a mild poison to a lethal one, now known as oxophenarsamine. (Later it would be shown that oxophenarsamine was a much more selective and desirable drug, and eventually it would replace salvarsan.) There were problems with side effects; also, at times, the syphilis had progressed so far that salvarsan was not effective as a cure. Ehrlich became concerned about the uncontrolled oxidation of salvarsan and decided that it had to be provided in sealed, single-dose ampoules, which would exclude oxygen. He also issued directions for its preparation in sterile water and its neutralization, and advised that intravenous injection must be carried out without delay. These directions were often not adhered to, and the resulting deaths attracted much unfavorable publicity. Moreover, salvarsan was not selective when given as a single dose; instead, treatment had to be spread out over several months. This meant that fewer than 25% of the patients ever completed the course of treatment. Despite these problems, salvarsan and its successor (neoarsphenamine or 914, a more soluble derivative of 606) remained the best available treatment for 40 years. Ehrlich shared the 1908 Nobel Prize with von Behring for their work on immunity; however, he regarded his greatest contribution to be the development of a "magic bullet," salvarsan. At the Nobel ceremony he modestly said, "My dear colleagues for seven years of misfortune I had one moment of good luck!"

The era of chemotherapy began with the work of Paul Ehrlich in his Frankfurt laboratory, and it continues to this day. The era of therapy by antibiotics—chemotherapeutic agents made by other microbes, not by chemical synthesis in a laboratory—began much later, with a chance discovery made by Alexander Fleming (1881 to 1955) at St. Mary's Hospital in London. In 1929 Fleming found that one of his Petri dishes had a lawn of streptococci with a zone of inhibition around a contaminating green mold, *Penicillium*. Although Fleming coined the term "antibiosis"—literally "against life"—for the phenomenon of bacterial growth inhibition, his subsequent work using "mold juice" produced inconclusive results and he discontinued further research. A decade later, a British pathology group at Oxford University, headed by a brash Australian, Howard Florey, and a temperamental Jewish refugee, Ernst Chain, took up the project. Despite working under wartime conditions, and with limited equipment and supplies, they eventually purified and tested the active material, named penicillin. Mice were infected with a fatal dose of streptococci; within hours the mice that had been given penicillin survived whereas the penicillin-deprived mice were dead. The curative possibilities for this wonder drug were obvious, and by the end of 1943 penicillin production was the second highest priority in the U.S. War Department. Because the British Medical Research Council took the position that patenting medicines was unethical, American companies patented the production techniques they had been using. Fleming, Florey, and Chain shared the 1945 Nobel Prize for their work on penicillin; however, accolades and media attention were disproportionately showered on Fleming. Regrettably, the contributions of Florey and Chain were soon forgotten and Fleming alone came to be considered responsible for penicillin. By the late 1940s, penicillin, now available in amounts suitable for the treatment of human disease, became the drug of choice for the treatment of syphilis and other bacterial diseases.

Only later would its mode of action be discovered: penicillin blocks the synthesis of the bacterial cell wall.

Resistance

Just as there is resiliency in the human species, there is also resiliency in other organisms that enable them to survive in the presence of drugs. This phenomenon is called drug resistance. How does resistance develop? Mutations may occur in the organism that permit its survival, and this capacity is passed on to its offspring. In its most basic form, this is survival of the fittest! The presence of the drug acts as a selective agent—a sieve, if you will—that culls out the sensitive organisms and allows only the resistant ones to pass through to the next generation. There are three basic mechanisms that allow microbes (and other organisms such as insects) to become drug resistant:

1. They become impermeable to the drug, or they pump the drug out of the cell so that toxic levels are not reached. This mechanism was first shown by Ehrlich when he found that some trypanosomes, which were resistant to trypan red, did not absorb the dye, whereas those that were susceptible stained red. (This observation clearly supports Ehrlich's principle that for a drug to be effective it must be chemically bound to the parasite.)

2. They develop an altered enzyme, which has a lower affinity for the drug.

3. They manufacture excessive amounts of enzyme, thus counteracting the drug, a phenomenon called gene amplification.

Drug resistance can develop without exposure to the drug, but once the drug is present, natural selection promotes the survival of resistant individuals. Drug resistance frequently blunts our ability to eradicate pathogens, and it can enhance virulence. However, not all microbes are able to become drug resistant. The genome of *T. pallidum* has remained relatively stable and constant; as a consequence, it is virtually the only pathogenic microbe that has remained as sensitive to penicillin as when penicillin was first introduced. On the other hand, the composition of its lipid outer surface, which is similar to that of human cells, and the uniquely exposed surface protein encoded by *tpp15* allow *T. pallidum* to evade immune control, leading to a chronic infection.

Syphilis and the Social Reformers

Although the advent of penicillin has helped to significantly reduce the incidence of syphilis, the disease has by no means been eradicated. A great deal of the failure can be attributed to social attitudes toward syphilis in particular and sexually transmitted disease in general. When salvarsan treatment became available in the early 1900s, syphilis became the focus of social reformers since to them the disease was a clear indicator of the breakdown of the home, morality, and marital fidelity and could be ascribed to promiscuity, prostitution, and immigration. So, they said, something had to be done about ridding society of the disease and its carriers.

Social reformers focused on the impact of sexually transmitted diseases on the family and the infected person—generally a man or woman who had strayed from the path of moral virtue and consorted with prostitutes or people of ill repute. Such individuals in turn infected the innocent spouse and the children. The moral code at the time (the Victorian era) held that sex within marriage was the only socially sanctioned sexual activity and that therefore the victims of syphilis "deserved" the disease and had brought it on themselves. The disease was considered a punishment for this kind of immoral sexual behavior. This is the same litany expressed about AIDS: that those who were infected deserved it because of their sexual activity whereas others were innocents who received it either from their infected mother at birth or via a contaminated blood transfusion. This attitude led public health officials in the early 1900s to spread the idea that simple contact was sufficient to spread syphilis, that one could get it from using a drinking cup; touching doorknobs, pencils, or pens; or sitting on a toilet seat. This was of course not true (because of the extreme fragility of the bacterium), but saying that so-called "innocent" transmission was possible allowed people to be socially "pure" even if they had contracted the disease. Physicians and politicians also used such information as a further tool to keep people from engaging in socially unacceptable behavior and as a means of reducing immigration into the United States (because immigrants were supposedly spreading syphilis in America either via prostitution or through casual contacts).

By the time of World War I, concern about syphilis had reached unprecedentedly high levels. The military draft and consequent medical examinations had revealed that 13% of those drafted were infected with either syphilis or gonorrhea, and this touched off a vigorous anti-VD (venereal disease) campaign. The focus of the campaign was on prostitutes, who were seen as the source of infection of military personnel. The idea was that, as one federal official stated, "To drain a red-light district and thereby to destroy the breeding ground of syphilis and gonorrhea is as logical as to drain a swamp thereby destroying the breeding place of malaria and yellow fever." As a result, many thousands of women near military training sites were quarantined during the war.

Those in charge of the VD campaign wanted not only to stop the spread of disease but also to change sexual and social behavior. It was not sufficient that soldiers should be cured of syphilis or that they should adopt precautions to avoid getting it whether they wanted to have sex or not. They were supposed to live a morally pure life and avoid extramarital sex under any circumstances, and in doing so they would avoid contracting a sexually transmitted disease. Latex condoms were available during the time of World War I, but the military authorities refused to provide them to the troops because it was believed that this would contribute to their moral decline and encourage sexual behavior. The treatment for syphilis was made painful and was intended to be a deterrent to contracting the disease. The U.S. Army also ruled that sexually transmitted diseases were injuries not incurred "in the line of duty," and so those who became infected lost their pay. The whole attitude was punitive: the infected GI had done something wrong, and the next best thing to telling them not to do it in the first place was to punish them afterward.

As one might expect, the campaign was unsuccessful. Rates of sexually transmitted diseases continued to stay at high levels or rise during and after WW I, and efforts to control syphilis lagged during the 1920s. In the 1930s, President Franklin D. Roosevelt appointed Thomas Parran as Surgeon General. Parran avoided taking a moral stance on disease and instead concentrated on how to eradicate it—an achievable goal, but one that was unpopular in some quarters. Parran developed a five-point program based on successful programs in Scandinavia. (i) Find the cases. To this end, diagnostic centers were established where there would be confidentiality and free testing to identify cases. (ii) Treat infected individuals promptly to prevent disease spread and to reduce virulence. (iii) Break the chain of transmission by tracing and treating sexual contacts. (iv) Institute mandatory premarital testing using the Wassermann test. (In order to get a marriage license, individuals would have to be Wassermann negative; while in principle this was a reasonable idea, Wassermann testing for premarital screening turned out not to be effective because of false positives [i.e., some individuals who did not have syphilis had a positive Wassermann test], and even a positive blood test did not indicate an active infection. Further, the tests were directed at the lowest-risk group, and this requirement assumes that those who wanted to marry would not do so if they tested positive and that nobody would acquire an infection after marriage. By the 1980s, mandatory premarital Wassermann testing was dropped.) (v) Institute a massive public education campaign to inform people about the disease, how it is contracted, its symptoms, and where to get treatment. The heart of the campaign was provision of information: people were not told what to do or what not to do, and no moral position was taken. The plan met with some opposition from the public and private sector since sexually transmitted disease was something polite people were not supposed to discuss in public. Indeed, when Parran was to make a radio broadcast about the topic on CBS in 1934, he was allowed to do so only if he did not use the word syphilis or gonorrhea. His will did eventually prevail, and the National Venereal Disease Control Act was passed in 1938. Congress allocated money for sexually transmitted disease clinics and testing, and treatment was provided to those who could not afford it.

During World War II, antisyphilis efforts once again intensified when it was found that ~5% of the recruits had syphilis. Unlike World War I, troops were provided with condoms, education, prophylaxis, and rapid treatment. If a GI did not report the disease, however, it was a court martial offense.

Today's Concerns

Syphilis is not uniformly distributed across the globe. According to a 2001 World Health Organization report there were over 12 million cases of syphilis worldwide in 1999. The vast majority of these cases occurred in the developing world, where as much as 10% of the population was infected: four million in sub-Saharan Africa, three million in Latin America and the Caribbean, and four million in South and Southeast Asia. Each year, 500,000 infants are born with congenital syphilis and there are another 500,000 miscarriages due to maternal syphilis. In Western Europe, syphilis prevalence has declined since the end of World War II, but the incidence remains at >5 cases per 100,000 population. Studies of pregnant women in Africa show rates of 17% in Cameroon, 8.4% in South Africa, and 6.7% in the Central African Republic. In South Pacific countries the rate is 8%, and in Morocco and Sudan the range is 2.4 to 4.0%. The incidence of syphilis is causally related to human immunodeficiency virus (HIV) transmission, with a two- to fivefold-increased risk of acquiring an HIV infection when syphilis is present.

In the United States, thanks to the introduction of antibiotics, the incidence of primary and secondary syphilis fell from 72 per 100,000 population in 1940 to 20 per 100,000 in 1990 to 3.2 per 100,000 in 1997, and to 2.7 per 100,000 by 2004. In 2004, public health officials in the United States reported over 7,980 cases of primary and secondary syphilis. Of these, 84% occurred in men. There were 353 cases of congenital syphilis in newborns. Untreated early syphilis during pregnancy results in perinatal death in up to 40% of fetuses; if acquired during the 4 years preceding pregnancy, it may lead to infection of the fetus in over 70% of cases. The rates of syphilis infection were highest in the South (3.6 cases per 100,000 population) and lowest in the Midwest (1.6 cases per 100,000 population). In 2004, 79% of 3,140 counties in the United States reported no cases of primary and secondary syphilis. The rate of syphilis was 5.6 times higher among African Americans than among non-Hispanic whites and higher among African American men (14.1 cases per 100,000) than women (4.3 cases per 100,000 population). The case rate for Hispanics was 3.2 per 100,000, that for whites was 1.6, that for Asians and Pacific Islanders was 1.2, and that for American Indian and Alaskan Natives was 3.2. In 2004, more than 60% of new infections occurred in men who have sex with men. The increase in this group is associated with coinfection with HIV, high-risk sexual behavior, and the use of drugs such as methamphetamines.

Primary prevention includes safer sexual practices and earlier identification and treatment. The current U.S. plan for achieving elimination of syphilis includes enhanced surveillance, strengthened community involvement and partnerships, rapid outbreak response, expanded clinical and laboratory services, and improved health promotion.

Control

Many of the attitudes of the public toward the AIDS epidemic (see p. 190) are reminiscent of the attitude toward syphilis a century ago. Although AIDS differs from syphilis in many respects, there are some important similarities. As before, there are two main theories on how a sexually transmitted disease should be fought. One, the moral approach, contends that the best way to prevent infection is to advocate a social and sexual ethic that makes it impossible to acquire an infection—sexual abstinence until marriage; this can allegedly be achieved through education and the suppression of prostitution. The other approach attempts to divorce itself from any particular judgment about sexual behavior and suggests that individuals should be provided with the means for protecting themselves from infection should they choose to engage in sexual behavior, and that treatment should be nonpunitive so as to encourage infected individuals to seek help.

Can syphilis be controlled? Yes. Because of its characteristics, it may be susceptible to control and perhaps even elimination. There are no animal reservoirs, the incubation period of 9 to 90 days allows for interruption of transmission by therapy of sexual contacts, and diagnosis is inexpensive and widely available. Further, it is treatable with a single shot of penicillin, and microbial resistance has not been reported. However, in some countries there remain barriers to control. Since the spread of an infectious disease depends on the average number of new cases generated by an infected individual and since individuals with syphilis are most contagious during the primary and secondary stages of the disease, which may last as long as 6 months to a year, the faster the infected individual seeks diagnosis and treatment, the faster the disease can be controlled; however, when diagnosis and treatment are delayed—and this depends on complex individual, social, cultural, and economic factors—transmission may continue. The available serological tests, though cheap and effective, are not perfect. They can give false-negative results and may be hard to interpret when other treponeme infections are present. Undoubtedly an important factor contributing to the persistence of syphilis is the disregard of those infected for others simply because the primary stage is painless and the signs of the secondary stage can be so variable. Then, too, there is the stigma associated with seeking care for a sexually transmitted disease. However, even when there is no stigma, there may be a lack of appropriate health care facilities. Although penicillin is cheap, it must be administered by injection, and those who dislike or fear the needle may avoid therapy. Another barrier to control can be the inability to identify and promptly treat potentially infected sexual partners. If a person who is diagnosed with syphilis cannot or will not identify his or her sexual contacts, one of Parran's critical elements for control will be lacking. The opportunity to interrupt disease transmission requires identifying, screening, and treating potentially infected individuals. If any of these elements are missed, control is impossible.

Consequences

Syphilis is more than a corrosive infectious disease. The history of syphilis is replete with discrimination against those with differing lifestyles and those who are marginalized in society—the urban poor, the uneducated, immigrants crowded into the cities, those with alternative lifestyles such as sexual orientation, and areas with inadequate access to health care. The terror of syphilis promoted the search for new drugs. However, drug treatments, even those that have been shown to be successful, cannot be relied on for complete eradication of a sexually transmitted disease such as syphilis. Syphilis galvanized public health authorities to stem its spread through education, treatment, avoidance of moral judgments, and attempts to blunt stigmatization. The incidence of syphilis, however, is a reflection of multiple factors: cultural beliefs and practices as well as economic and political forces. Its control will require comprehensive programs for diagnosis as well as treatment, but it is critical that these be suited to the social and cultural dynamics of this sexually transmitted disease.

7

Tuberculosis: the People's Plague

Violetta in Giuseppe Verdi's opera "La Traviata" (1853) and Mimi in Giacomo Puccini's opera "La Boheme" (1895) are young, tall, thin, and pale-faced with cherry-red lips and flushed cheeks, and their voices are like those of the nightingale. But Mimi and Violetta are also mysteriously ill with a wasting disease called consumption (from the Latin "con," meaning "completely," and "sumere," meaning "to take up"). To those living in the 19th century it was natural to link artistic talent to consumption, and Verdi and Puccini were well acquainted with this connection. Indeed, at the time it was believed that the poisons of consumption stimulated mental activity and artistic talent.

Consumption is more commonly called tuberculosis (TB). In *Illness as a Metaphor*, Susan Sontag wrote, "TB was—still is—thought to produce spells of euphoria, increased appetite, exacerbated sexual desire . . . Having TB was imagined to be an aphrodisiac, and to confer extraordinary powers of seduction." In the 1800s, when epidemic TB reached its peak in Western Europe, infected persons were considered beautiful and erotic, with their extreme thinness, long neck and hands, shining eyes, pale skin, and red cheeks. Yet such "beauty" had its price: a painful death by drowning in one's own blood. Because it was neither recognized nor understood that TB was a chronic infectious disease, it was romanticized, mythologized, and regarded as a spiritualizing force.

The operas "La Traviata" and "La Boheme" were both based on Alexandre Dumas' 1848 semi-autobiographical novel *The Woman of the Camellias* (made into the classic 1936 movie "Camille" with Greta Garbo and Robert Taylor). In the Dumas work, the heroine (a courtesan) coughs blood onto her white handkerchief, recalling the red and white colors of the camellia flower and symbolically representing her sexual availability at various times during the menstrual cycle. (The spitting up of blood was sanitized in the operatic versions.) A more accurate and less romantic description of the consumptive included incessant coughing, which made talking and eating almost impossible and made breathing painful; loss of weight, which prevented walking; and severe pain, which required opium and whisky for control. By the time of death, emaciation was so complete that the individual resembled a cadaver. The romantic notion of TB in "La Traviata" (literally "the fallen woman") can be explained in part by the fact that the opera was composed before the cause of TB was discovered in 1882. The themes of artistic genius and eroticism persist in "La Boheme" despite its later date, largely because "consumptive decline" was considered to be due to a hereditary predisposition or specific living habits such as the debauched bohemian lifestyle, poverty, or sexual promiscuity. This romanticized and morbid fascination with consumption is entirely without scientific basis, yet the linkage between a disease, creativity, and eroticism persists: in the 1996 rock musical "Rent," Puccini's beloved tubercular heroine and her fellow bohemians are transplanted to New York City, and despite giving them AIDS, rats, and roaches, composer Jonathan Larson passionately describes their hopes, losses, striving, death, and climactic resurrection.

Origins

TB is an ancient disease that has plagued humans throughout recorded history and even earlier. TB of the lungs is the most familiar form of the disease, giving rise to the slang word "lunger." When localized to the lungs, TB can run an acute course, causing extensive destruction of lung tissue in a few months—so-called galloping consumption. It can also wax and wane, with periods of remission, and in this manifestation it is sometimes mistaken for chronic bronchitis with spitting up of blood. In 1839 Harriet Webster, the daughter of Noah Webster (author of Webster's dictionary), wrote, "I began to cough and the first mouthful I knew from the look and feeling was blood . . . I concluded to lay still and try what perfect quiet could do—swallowed two mouthfuls of blood and became convinced that if I could keep from further coughing I should be able to wait until morning without disturbing anyone. As soon as morning arrived, I looked at the contents of my cup. Alas, my fears were realized." She died 5 years later.

TB can affect organs other than the lungs, including the intestine and larynx; sometimes the lymph nodes in the neck are affected, producing a swelling called scrofula (scrofula comes from the Latin word for "pig" because the shape of the swollen neck looks like a little pig). TB can also result in fusion of the vertebrae and deformation of the spine, called Pott's disease after Sir Percival Pott, who described the condition in 1779. This may lead to a hunchback and may also affect the skin and the kidneys. TB of the adrenal cortex destroys adrenal function and results in Addison's disease; this is probably what killed Jane Austen.

Evidence of TB is found in bony remains that pre-date human writing. Pott's disease has been described from Egyptian mummies dating from 3700 to 1000 BC. One of these mummies, from the XXI Dynasty (1000 BC), discovered near the city of Thebes in 1891, is that of a priest who died with extensive destruction of the bones of the spine. Curiously, there is no evidence for lung disease in any of the mummies from this period; however, in the tomb of a high priest of Ramses II the body of a small boy has been found whose lungs had been stored, and these show the telltale signs of pulmonary TB. The burial site has been dated to 400 to 1000 BC. Thus, it appears that tubercular disease of the lungs is more recent than that of the bones. Based on this, it has been suggested that human TB evolved from a disease of cattle after they were domesticated between 8000 and 4000 BC. Before this time, the epidemic form of TB as we know it today did not occur. It is thought that TB then spread to the Middle East, Greece, and India via nomadic tribes (Indo-Europeans) of milk-drinking herdsmen who had migrated from the forests of central and eastern Europe in around 1500 BC. A clay tablet in the library of the Assyrian King Ashurbanipal (668 to 626 BC) describes the disease: the patient coughs frequently, and his sputum is thick and sometimes contains blood. His breathing is like a flute. His skin is cold. The Greek physician Hippocrates (460 to ca. 375 BC) called the disease "phthisis," meaning "to waste," and noted that the individual was emaciated and debilitated and had red cheeks and that the disease was a cause of great suffering and death. Although Hippocrates believed that the disease was due to evil air, he did not consider it contagious. Aristotle (384 to 322 BC), however, suggested that it might be contagious and may be due to "bad and heavy breath." However, by the time of Galen (AD 129 to ca. 200), the theory of contagion of phthisis came to be accepted in the Roman Empire, but the contagious agent was not found.

During the Middle Ages (AD 500 to 1500), a feudal system developed in Europe, where a small elite (the nobility) ruled the rest of society, their subjects. Royalty claimed that their rights to rule and their talents were of divine origin, and they publicized this through claims of royal supernatural powers to heal disease, specifically scrofula. Kings and queens were claimed to be able to heal those afflicted with scrofula by a simple touching. Clovis of France (reigned 481 to 511) and Edward the Confessor of England (reigned 1042 to 1066) were supposedly the first kings endowed with this gift. Edward I (reigned 1272 to 1307) touched 533 of his subjects in one month, Philip VI of Valois (reigned 1328 to 1350) touched 1,500 in a single ceremony, and Charles II (reigned 1660 to 1682) touched 92,102 people during his reign. On 14 June 1775, Louis XVI "ritually touched 2,400 stinking sufferers from scrofula."

During the ritual ceremony, the king or queen touched the sufferer, made the sign of the cross, and provided the afflicted with a gold coin. In England this practice was known as the King's Evil or the Royal Touching, and it persisted until the early 18th century. One of the last sufferers of scrofula to be touched was the English writer, critic, and lexicographer Dr. Samuel Johnson, who was infected with tuberculosis by his wet nurse. Johnson was brought to Queen Anne (reigned 1702 to 1714) at the age of 2 years and touched; from all evidence, it appears that the touching did not cure him. However, the crowd of sufferers anxious to be cured rushed to be touched, and several were trampled to death on this last occasion. Today we know that royal touching had nothing whatever to do with curing scrofula; cure occurred in some cases because of the natural remission of the disease.

The word "tuberculosis" refers to the fact that the disease causes characteristic small knots or nodules called "tubercles" in the lungs. Franciscus Sylvius first described these in 1650; he also described their evolution into what he called lung ulcers. However, almost all of the great pathologists of his time believed that the disease was due to tumors or abnormal glands rather than an infection. The first credible speculation on the infectious nature of TB was that of Benjamin Marten, who in 1722 proposed that the cause was an "animalcule or their seed" transmitted by the "Breath (a consumptive) emits from his Lungs that may be caught by a sound Person." However, it would take 100 years before that "animalcule" was identified.

In the 18th century, the present epidemic wave of TB began in England, reaching its peak in about 1780. It was estimated that 20% of all deaths in England and Wales were due to consumption at that time. From there it spread to the rest of Western Europe, reaching a peak in the 1800s, when "La Traviata" and "La Boheme" were written. Peaks of TB incidence occurred in Eastern Europe between 1875 and 1880, and by 1900 the infection had reached North America. The cause of this rise in TB incidence may have been a demographic shift from rural to urban living as well as the creation of "town dairies." These dairies were wooden buildings in the town center which housed dairy cows that had formerly been pastured; now the tubercular cows were within the town, which provided ideal circumstances for animal-to-animal as well as animal-to-human (zoonotic) transmission of TB. The result was a sharp rise of scrofula in the 17th century. Later, with a resurgence of trade, the walled towns of England continued to provide the means of human-to-human transmission. This was especially so when the textile industry became mechanized; this development led to a shift from a rural cottage industry to the more urban riverside sites where waterpower was available. Towns grew to become cities; as peasants streamed into these urban centers, people were more and more crowded. These conditions are vividly shown in the paintings of Pieter Bruegel (1525 to 1569) and William Hogarth (1697 to 1764). Further, the practice in England of taxing a building partly based on its number of windows tended to affect building design to minimize the number of windows, which enhanced the rebreathing of exhaled air of those living and working in such crowded, airless rooms. The increased density of people provided ideal conditions for the aerial transmission of TB. By the 19th century, epidemic TB had raged for more than two centuries, and some feared that it might bring about the collapse of industrialized Europe as well as the end of all civilization. An early 20th century journalist described it: "Tuberculosis is a Plague in disguise. Its ravages are insidious, slow. They have never roused people to great, sweeping action. The Black Plague in London is ever remembered with horror. It lived one year; it killed fifty thousand. The Plague Consumption kills this year in Europe over a million; and this has been going on not for one year but for centuries. It is the Plague of all plagues—both in age and power—insidious, steady, unceasing."

To those living during the Victorian Age (1837 to 1901), when the British Empire was at its height, TB was romantic and attractive since it produced no obvious repulsive lesions. To the Victorians, the blood in the sputum blended metaphorically with menstrual blood, and so in a strange way sickness and death were blended with eroticism and procreation. There is an eminent gallery of victims of tuberculosis including Baruch Spinoza (1633 to 1677), Johann Wolfgang Goethe (1749 to 1832), Friedrich

Schiller (1759 to 1805), Fyodor Dostoevsky (1821 to 1881), Anton Chekhov (1860 to 1904), Sir Walter Scott (1771 to 1832), D. H. Lawrence (1885 to 1930), Percy Bysshe Shelley (1792 to 1822), John Keats (1795 to 1821), Alexander Pope (1688 to 1744), Samuel Johnson (1709 to 1784), Jean Antoine Watteau (1684 to 1721), Niccolo Paganini (1782 to 1840), Elizabeth Barrett Browning (1806 to 1861), Igor Stravinsky (1882 to 1971), Robert Louis Stevenson (1850 to 1894), Edgar Alan Poe (1809 to 1849), Franz Kafka (1883 to 1924), Amadeo Modigliani (1884 to 1920), Frederic Chopin (1810 to 1849), Henry David Thoreau (1817 to 1862), George Orwell (1903 to 1950), Eleanor Roosevelt (1884 to 1962), and Vivien Leigh (1913 to 1967). Keats died at age 21, and all six children of Reverend Brontë and his wife Maria (including Emily and Charlotte), as well as Shelley, died of "galloping consumption" before the age of 40. Some, looking at this list of writers, composers, and artists, believed that TB sparked genius; Arthur Fishberg wrote, "Tuberculosis patients particularly young talented individuals . . . display enormous intellectual capacity of the creative kind. Especially is this to be noted in those who are of artistic temperament, or who have a talent for imaginative writing. They are in a constant state of nervous irritability, but despite the fact that it hurts their physical condition, they keep on working and produce their best works." However, there is little scientific evidence to show that tuberculosis had any real effect on the brain or on creativity.

The deaths of Vivien Leigh and Eleanor Roosevelt are less romantic and more tragic because they occurred after the disease agent had been identified and drugs for the cure were available. The actress Vivien Leigh developed TB in 1945, was hospitalized for a brief time, and recovered, but her physician recommended that she seek further treatment and hospitalization. She refused. Although antibiotics became available to treat TB in the 1940s and 1950s, she again refused to avail herself of these cures. She died of TB at age 54. Eleanor Roosevelt developed active TB at age 12, and then the disease went into remission. In the last 2 years of her life, her health began to deteriorate and she developed miliary TB (so called because the small tubercles in the lungs look like millet seeds, and these spread throughout the body via the bloodstream). It was too late for treatment, and she died of disseminated TB at age 75.

The bacteria that cause TB are called mycobacteria. Their free-living relatives inhabit the soil and water, where they fix nitrogen and degrade organic materials. Mycobacteria have a protective cell wall that is rich in unusual waxy lipids such as mycolic acid and polysaccharides such as lipoarabomanan and aribinogalactan. *Mycobacterium tuberculosis* and *M. avium* are human pathogens that cause lung disease. *M. avium* causes opportunistic infections (see p. 178) in immunocompromised people (e.g., those with AIDS); its symptoms can include weight loss, fevers, chills, night sweats, abdominal pains, diarrhea, and overall weakness. Closely related to *M. tuberculosis* is a parasite of cattle, *M. bovis*. Although *M. bovis* can infect humans, it does so infrequently and with great difficulty. *M. bovis* grows under conditions where the oxygen levels are low. When it does infect people, it is not associated with lung disease. *M. tuberculosis*, in contrast, grows best when oxygen is plentiful; it is associated with pulmonary TB, probably because the lungs contain high levels of oxygen. However, TB of the spine (Pott's disease) is associated with *M. bovis* and results from a blood infection that spreads to the spine via the lymph vessels. It has been hypothesized that *M. bovis* arose from soil bacteria and that humans first became infected with *M. bovis* by drinking milk. *M. tuberculosis*, on the other hand, is specific to humans and spreads from person to person through airborne droplets of saliva and mucus. Genetically, *M. bovis* and *M. tuberculosis* are >99.5% identical, so the differences in their pathogenic nature are still to be explained.

Some contend that TB was brought to the Americas by European explorers and settlers; however, this appears not be true. An Inca mummy of an 8-year-old boy who lived about AD 700—more than 700 years before Columbus and his crew arrived in the Americas—shows clear evidence of Pott's disease, and the lesions in the spine contain bacteria, most probably *M. bovis*. Domestic herds of cattle as well as wild herbivores probably served as the source of infection. In 1999 investigators found *M. bovis* in bone tissue removed from a 17,000-year-old bison that had fallen to its death in the Natural Trap Cave in Wyoming, providing clear evidence that TB was already present in prehistoric America, waiting for new human hosts. With urbanization, immigration of infected individuals, higher population densities, and poor hygiene, the conditions for the spread of TB in the New World were favorable. Its first peak was in the early 19th century, and TB was especially prevalent in the cities along the Atlantic coast. Mark Caldwell, in his book *The Last Crusade*, described it: "Though its crowds and bad air made the city a crucible of TB, it was also the place where the poor congregated, where oppressive economic conditions prevailed, where filth and ugliness constantly assaulted the senses." In 1804 in New York City, a quarter of all deaths were due to consumption, and between 1812 and 1821, Boston had a similar fatality rate. TB did not affect all segments of the U.S. population equally. In 1850, African Americans in Baltimore and New York City had higher death rates than did whites. In Baltimore, the number of female deaths was twice that of males in those older than 15 years, but in New York and London it was the reverse. The reasons for this are not clear. TB was not strictly an urban disease; it was also present in rural areas. The critical element was found not to be the total population but the size of the household. In colonial America in the 18th century, irrespective of the size of the town or city, a typical dwelling had 7 to 10 inhabitants. Such crowding facilitated household transmission of the disease. Inefficient heating of the home usually led to sealing of the windows and doors in winter to keep out the cold, and so transmission was enhanced. Furthermore, behavioral patterns also contributed to the spread of TB: caretakers of the sick frequently slept in the same bed as their patients, and physicians advised against the opening of windows in rooms with patients who were sick with consumption.

Inadequate ventilation was also a contributing factor in the spread of urban tuberculosis. For example, during the 19th century most tenements were built with little concern for proper ventilation. As more and more impoverished immigrants arrived in the United States (especially during the 1830s and 1840s), they were forced to live in crowded conditions in these miserable tenements. In 1800 a Boston apartment typically contained an average of 8 people, and by 1845 the figure was over 10. The generally filthy conditions as well as lack of ventilation characteristic of tenement housing surely played a key role in the rise in the incidence of TB in the United States.

In the urban centers of New York and Boston, consumption came to be regarded as "a Jewish disease" or the "tailors' disease" because so many of the young immigrant Jews who were consumptives earned their living cutting, sewing, and stitching in the garment industry. Although TB was peculiar neither to Jews nor to those who worked in tailoring, these groups continued to be stigmatized as carriers of tuberculosis. During the 1920s, American nativists became concerned about the millions of emaciated and undersized Jewish immigrants who were lacking in the physical vitality so characteristic of the sturdy and robust Anglo-American stock and who were now pouring into the cities along the Atlantic seaboard. It was contended that because Jews were highly susceptible to TB, these sickly Jews not only were racially inferior but also would soon become public charges. Data gathered by the leaders of the Jewish community refuted the claim that Jews were especially prone to TB. Although the rates of consumption among Jews in both Europe and the United States were demonstrated to be lower than those among non-Jews, this did little to change the mind of the public. In effect, TB was used as a tool of anti-Semitism that justified the ostracism and persecution of Jews.

In the U.S. penitentiaries, approximately 10 to 12% of white inmates died of TB between 1829 and 1845, whereas during this time the annual mortality due to TB in eastern cities of the United States was less than 0.5%. Even today there is a higher incidence of TB in prisons. The increased incidence is due to several factors: the prevalence is higher among new prisoners than in the general population because there is a weighting of prisoners toward the lower end of the socioeconomic scale, close living arrangements make transmission more likely, and prisoners are at a higher risk for TB due to having a higher incidence of HIV infection.

In the 1900s, poorer people tended to have a higher mortality; the greatest number of deaths occurred between the ages of 15 and 45 years but there was also a minor peak between 5 and 10 years of age. American Indians, who were highly susceptible to TB, were virtually killed off by being herded together on reservations. Indeed, between 1911 and 1920, 26 to 35% of all deaths occurring among American Indians were due to TB, and an examination of 600,000 American Indians during 1920 showed that 36% had TB. TB was virtually unknown in sub-Saharan Africa until the beginning of the 20th century and was not found in areas where there had been little or no European immigration.

Discovery

In 1865 a French military physician, Jean-Antoine Villemin (1827 to 1892), succeeded in transmitting tuberculosis to rabbits. Villemin recovered pus from the lung cavity of a tuberculous patient who had died 33 h earlier and injected it under the skin of two healthy rabbits. Two other rabbits served as controls and were injected with tissue fluid from a burn blister. Some months later, when the rabbits were examined, TB was found in the lungs and the lymph nodes of only those that had received the pus. Although it would seem that Villemin's demonstration of transmission would be proof enough of the contagious nature of TB, his announcement before the prestigious French Academy of Medicine was greeted with derision. His severest critic was Hermann Pidoux, who said that consumption in the poor was due to conditions of poverty including overwork, malnutrition, unsanitary housing, and other deprivations; when consumption developed among the rich, Pidoux claimed it was brought about by their overindulgence in their wealth, laziness, flabbiness, overeating, excessive ambition, and habits of luxury. In both cases the result was, in Pidoux's words, "organic depletion." Such blind rejection of the infectious nature of TB remained until 24 March 1882, when the German physician Robert Koch made a presentation before a skeptical audience of Germany's most prestigious scientific group, the Berlin Physiological Society, in which he claimed to have discovered the microbe of tuberculosis. Koch's announcement not only resolved the contagion argument but also resulted in a shift of scientific preeminence from France (and Louis Pasteur) to Germany. This shift took place for several reasons, including differences in the style of science. French scientists relied on intellectual deduction and tended to proclaim general laws from a limited set of experiments, whereas German scientists were more methodical, their approach was more reality-based, and they relied on repetitive experiments. In addition, after the defeat of the French in the Franco-Prussian War (1870 to 1871), the German economy and its science flourished while post-Napoleonic France declined both in its wealth and in its scientific glory.

Robert Koch, born in Clausthal in the Harz Mountains of Germany in 1843, received his medical education at the University of Göttingen and then trained with the eminent pathologist Rudolph Virchow in Berlin. After the Franco-Prussian War he settled in the small farming town of Wollstein, where he had a medical practice and was the Sanitary Officer. The disease anthrax was a problem among the sheep (and sometimes humans) in and around Wollstein, so Koch, equipped with a new microscope his wife had given him for his birthday, began to examine the carcasses of sheep. Soon (in 1881) he was able to isolate and characterize the germ of anthrax, a rod-shaped bacterium. Koch presented his findings at the International Medical Congress in London in 1881. After returning from the meeting in London, he was determined to find the microbe that caused TB. Within a year he had succeeded. He wrote, "The aim of the study had to be directed first towards the demonstration of some kind of parasitic forms, which are foreign to the body and which might possibly be interpreted as the cause of the disease . . . The objects for study are prepared in the usual fashion for the examination of pathogenic bacteria . . . the coverslips are then covered with a concentrated solution of methylene blue . . . and when the coverslips are removed . . . the smear looks dark blue and is much overstained, but upon treatment with Bismarck brown stain the blue color disappears and the specimen appears faintly brown. Under the microscope all constituents of animal tissue, particularly the nuclei and their disintegration products, appear brown, with the tubercle bacilli, however, beautifully blue."

Koch also carefully and precisely set the criteria (called Koch's postulates) necessary for a microbe to be the causative agent of a disease:

> ". . . in all tuberculous affections of man and animals, there occurs constantly those bacilli which I have designated tubercle bacilli and which are distinguishable from all other microorganisms by characteristic properties. However, from the mere coincidental relation of these tuberculous affections and bacilli it may not be concluded that these two phenomena have a causal relation, notwithstanding the not inconsiderable degree of likelihood for this assumption that is derivable from the act that the bacilli occur by preference where tuberculous processes are incipient or progressing, and that they disappear where the disease comes to a standstill. To prove that tuberculosis is a parasitic disease, that it is caused by invasion of bacilli and that it is conditioned primarily by the growth and multiplication of the bacilli, it was necessary to isolate the bacilli from the body, to grow them in pure culture until they were freed from any disease-product of the animal organism which might adhere to them; and by administering the isolated bacilli to animals, to reproduce the same morbid condition which, as known, is obtained by inoculation with spontaneously developed tuberculous material."

Koch's identification of the "germ" of TB, *M. tuberculosis,* was not so simple: the germ is colorless and unusually difficult to stain because of its waxy cell wall, and therefore it cannot be easily seen under the light microscope. Rendering it visible required heating and a special aniline dye (methylene blue). Later, the staining technique was improved by Paul Ehrlich, who had heard Koch speak in 1882; Ehrlich found that the bacteria would retain the red dye color if they were stained with fuchsin followed by an acid wash (i.e., they were acid-fast). Koch also devised a method for growing them outside the body (by culturing them in test tubes containing coagulated serum as a nutrient source) and was able to use such pure cultures of tubercle bacteria to infect guinea pigs. It was methodical Koch, the German, who, although he validated the imaginative approach of the Frenchman, Villemin, ignored his findings and thus relegated him to obscurity. In this case, scientific credit was given to the person who convinced the world, not the one to whom the idea first occurred.

In 1890, under pressure from the Prussian government and its leader Otto von Bismarck, Koch forsook careful experimentation and prematurely announced that he had discovered a protective substance made from an extract of the bacillus called tuberculin (today known as purified protein derivative [PPD]). When injected into animals, tuberculin produced fever, malaise, and signs of illness. Koch believed that tuberculin sensitized the animal and effected a cure. It was soon found that it did not. Indeed, instead of being curative, tuberculin turned out to be dangerous or sometimes lethal. Tubercle bacilli sensitize the body to tuberculin, so that when tuberculin is injected in sufficient quantities into a tuberculous individual, it can kill via a delayed hypersensitivity reaction. Yet Koch's tuberculin did serve a practical purpose as a diagnostic test. Even today it remains one of the most useful methods for determining previous exposure to TB. The tuberculin skin test is positive in 15% of adults living in the United States, but a positive test does not indicate whether the disease is active. Indeed, only 10% of people exposed to *M. tuberculosis* develop active disease.

Koch was also able to show, as had Villemin a decade earlier, that bovine TB could be infectious for humans, principally via the milk. At this time, when infected milk was recognized as a source of human disease, the tuberculin skin test became useful as a means of screening infected cows. Cows that had a positive test result were killed. This meant that valuable animals would be taken out of production, and the losses could be devastating to farmers who had more than a few tuberculin-positive animals. However, slaughter was unnecessary since pasteurization of milk rendered it noninfective and protected those who drank such milk. Opposition to this slaughter-based control was vehement, and sometimes militia protection was needed to provide for the safety of the veterinarians who were involved in the testing in rural dairies where pasteurization was not employed. In 1932 there was a "Cow War" in Iowa as 400 angry farmers destroyed the veterinarians' car; ultimately, martial law had to be declared. By the end of the 1930s, more than 95% of counties in the United States were declared to be free of infected cattle and the threat of milk-borne TB faded away. However, the problem has not disappeared completely since bovine TB continues in other countries including Mexico.

The Disease

Charles Dickens, in his 1879 novel *Nicholas Nickelby*, described tuberculosis:

> "There is a dread disease, which so prepares its victims, as it were, for death . . . a dread disease. In which the struggle between soul and body is so

> gradual, quiet, solemn, and the results so sure, that day by day, and grain by grain, the mortal part wastes and withers away, so that the spirit grows light . . . a disease in which death and life are so strangely blended that death takes the glow and hue of life, and life the gaunt and grisly form of death—a disease which medicine never cured, wealth warded off, or poverty could boast exemption from—which sometimes moves in giant strides, or sometimes at a tardy sluggish pace, but slow or quick, is ever sure and certain."

Dickens could characterize TB, but he did not know that the principal risk behavior for contracting it was breathing. Persons with TB of the lungs may infect others through airborne transmission: coughing, sneezing, and speaking. A sneeze might contain an aerosol with a million microscopic (~10-μm) drops that evaporate slowly, producing floating droplet nuclei. TB bacilli, slightly bent microscopic rods ~2 to 4 μm long and 0.3 μm wide and enclosed within droplet nuclei, move from place to place by riding on the gentle air currents. The bacteria-laden droplets can easily be inhaled. Tubercle bacilli are rather robust and can survive in moist sputum for 6 to 8 months. Inhalation is the major route of infection. Oral infection, through eating or drinking, is less efficient because the bacteria rarely survive passage through the acid-containing stomach. Infection may result if as few as five bacteria reach the grape-like clusters of thin-walled air sacs (alveoli) of the lungs. In the United States and Europe, the lung is the primary site of infection in 80 to 85% of patients.

Once the bacteria are in the air sacs, they are engulfed by macrophages, and within the macrophage they are transported to other parts of the body by the lymph channels. For the first few weeks within the tissues of a susceptible individual (one who has not been tuberculous before), the bacteria multiply slowly, dividing once every 15 to 24 h. At first there is little damage or reaction, but after several more weeks of microbial multiplication, either at the initial site or at a more distant site, there is an inflammatory response; this becomes more intense, and fluid (lymph) leaks into the region. The site becomes infiltrated with fiber-secreting cells called fibroblasts that surround and wall off the free and macrophage-enclosed bacilli. The resulting capsule grows, pushing aside normal tissue, and produces the characteristic TB lesions, i.e., tubercles. In 90% of patients the disease progresses no further; however, in the other 10%, the tubercles break open, and when the small blood vessels that surround the air sacs are eroded, the bacilli can disseminate via the bloodstream; hemorrhages also occur, the bronchi become irritated, there is coughing, the sputum is streaked with blood, fluid fills the lungs, and breathing becomes more difficult. At this stage, acid-fast *M. tuberculosis* can be found in the sputum. Increasing destruction of more lung tissue produces a cheese-like consistency in which the bacteria survive but cannot multiply because of low levels of oxygen and acidity. Coughing, pallor, spitting of blood, night sweats, and painful breathing signify that the disease is active. Chest X rays show fluid and tubercles in the lungs; sounds of gurgling and slush can be heard using a stethoscope placed on the chest. Unfortunately, continued inflammation results in liquefaction of the lung tissue, and this oxygen-rich environment provides a rich growth medium; in some cases there can be more than 10 billion bacilli per ml. (It is a bitter irony that the tubercle bacillus flourishes in the apex of the lung where oxygen is plentiful and that fresh air, rich in oxygen, was believed to be curative for consumptives.) In time, the softened and liquefied lung contents are forced out of the lesion into the blood vessels, and this is how the disease spreads to other parts of the body. The fatality rate of untreated TB can be between 40 and 60%.

Tuberculosis is a self-limiting infection that often goes unnoticed—its symptoms are similar to the common cold, and usually there is little impairment of lung function. If there is a protective immune response, which occurs in ~90% of the cases, the disease may progress no further because calcification of the tubercles takes place. This condition, called latent TB, renders the individual noninfectious. However, in ~10% of cases a latent infection may become active—a risk that is greatest within 2 years of the primary infection and can be considerably higher in individuals with an impaired immune system, especially those with HIV infections. Persons with latent TB should be given treatment to prevent the infection from becoming active (see p. 125).

Reactivation of infection occurs because some of the tubercle bacteria, even those within macrophages, are not killed. For tuberculosis it is important to distinguish between infection and disease. In the vast majority of cases when the bacteria are inhaled, the bacilli are killed by macrophages or are localized and grow slowly within tubercles. Although *M. tuberculosis* can grow and divide outside of the cell, it survives predominantly within macrophages; it can live within the macrophage because its waxy-lipid coat makes it impervious to the killing mechanisms of the macrophage. Despite the formation of antibodies during a TB infection, these are unable to limit the disease because the bacilli remain hidden within the macrophage. Cell-mediated immunity, especially that involving CD4 T helper cells (lymphocytes), is critical for disease arrest. The CD4 T cells produce cytokines, especially gamma interferon and tumor necrosis factor, that activate macrophages to kill *M. tuberculosis* or limit its growth. The lesion becomes infiltrated with lymphocytes and macrophages, and a delayed hypersensitivity reaction, similar to that experienced with a bee sting, occurs. However, in tuberculosis, cell-mediated immunity is a two-edged sword—it is required for protection but is also involved in tissue damage and disease. In addition to the CD4 T helper lymphocytes, CD8 T cells (killer lymphocytes) participate in the immune response; these cells produce interferon gamma and tumor necrosis factor that activate macrophages, and in addition the cells punch holes in the membrane of the macrophage by releasing a molecule called perforin. Many macrophages are also killed by tuberculin-like products, causing them to release reactive oxygen species (hydrogen peroxide, hydroxyl radicals, and superoxide) and protein-degrading enzymes (proteases) that are detrimental to the host tissues. The products of the surviving tubercle bacilli are able to activate suppressor T cells that depress the delayed hypersensitivity reaction as well as cell-mediated macrophage killing, allowing the bacteria to spread to other tissues and the inflammatory cycle and tissue destruction to begin anew. It is ironic that the macrophages that are able to deal with some of the tubercle bacilli by using their cell-killing mechanisms can damage the very lung tissues they are designed to protect.

Prelude to Prevention

TB is a corrosive disease, and most of its classic symptoms—pallor, coughing, spitting up of blood, weakness, and emaciation—were indicative of an advanced state of the disease. Much of the improvement in our understanding of prevention from and treatment for TB has involved diagnosis and the development of methods to limit the spread of the "germ" both within and from outside the body.

Tuberculin, discovered by Koch in 1890, turned out not to be a cure for TB; however, in the control of TB it serves as the gold standard for diagnosis. In the tuberculin test a small amount of tuberculin (or PPD) is injected under the skin of the forearm, and within 48 to 72 h after injection an inflamed area develops if the person has been exposed to TB. The redness persists for up to 7 days. A positive skin test does not indicate an active infectious case of TB but merely shows that the person has been exposed. Another method of diagnosis is the one first demonstrated by Koch: culture of the bacteria in the laboratory followed by staining for the tubercle bacilli.

In 1895, Wilhelm Roentgen (1845 to 1923) discovered X rays. X-ray photography, or radiography, made visible the tubercular lesions caused by the disease long before its symptoms became noticeable; this would allow for treatment of the disease at a much earlier stage. However, X-ray photography did not become a reliable diagnostic tool until the 1920s, and even then treatments for TB remained less than satisfactory.

The French physician René Laennec developed another diagnostic tool, the stethoscope, in 1816. Laennec was asked to examine a rather overweight woman. Realizing that he could not directly tap on her chest to determine whether there was fluid in her lungs, he recalled from his boyhood that one could hear the scratch of a pin when the ear is placed in contact with one end of a wooden cylinder and the other end is scratched. In an imaginative stroke, he took a paper notebook, rolled it up tight, applied it to the woman's chest, and listened. Much to his surprise, he clearly heard the sounds of the woman's heart and lungs. Further experimentation eventually led him to develop the first stethoscope, a hollow wooden tube about 1 ft long. Laennec's stethoscope has been improved on, but it continues to provide the physician with an acoustic picture of the conditions within the lungs. However, it cannot specifically identify the specific agent responsible for any abnormal chest sounds. Based on his clinical experience, Laennec theorized that phthisis and scrofula as well as miliary TB were different forms of the same disease; this idea was opposed by German scientists, who were convinced that a specific clinical picture had to be the result of a single infectious agent. However, when Villemin demonstrated transmission of TB to rabbits and guinea pigs, it became clear that Laennec was correct. Laennec, who devoted his entire life to the study of TB, died of it in 1826.

"Catching" TB

By the 1940s, before antibiotics were introduced and strict public health measures were instituted, the lethal incidence of epidemic TB was declining. Why? Some suggested that it was due to the emergence of less virulent strains of the mycobacterium, whereas others proposed that it was due to an increase in the resistance of the human population. Still others contended that public health measures such as segregation of infected persons in sanatoria as well as forbidding the practice of spitting in public led to a decline in TB. Higher standards of hygiene including cleaner cities, destruction of tubercular cows, pasteurization of milk, and hand washing all led to reduced transmission; better nutrition and a higher standard of living also helped. A recent mathematical model suggests that all of these were contributory factors and that the decline was due to the natural behavior of this epidemic. Blower and colleagues have calculated that the great epidemics of TB did not occur in Europe until the 17th and 18th centuries (and somewhat later in the Americas) because before this time the population size was too small for effective transmission to occur. However, with urbanization and crowding as well as industrialization and abundant poverty and malnutrition, the population size exceeded the threshold for epidemic spread. Indeed, Blower and colleagues estimate that the number of secondary cases from a single infected source might have been as great as 10! When the course of the epidemic was fast paced, the disease was found primarily in younger individuals, but as the epidemic matured there was a shift toward infections in older individuals; these were the result of reactivation. According to this model, TB epidemics operate on an extremely long timescale—100 to 200 years—and the decline phase may take at least 30 years. These slow dynamics are "the result of the gradual development of a large pool of latently infected noninfectious individuals. Some of these individuals, however, gradually develop disease (and become infectious) over their lifetime." Before 1985 the highest rates of TB in the United States were found in individuals 45 to 64 year of age. This is characteristic of a mature epidemic where infection is the result of reactivation. Such a natural decline produced endemic equilibrium levels that gave the impression that TB was disappearing as a result of public health interventions. However, since 1985 there has been an increase in the incidence of TB, and the infections occur in a younger age group (as was seen in the heroines of the "Lady of the Camellias," "La Boheme," and "La Traviata"). This age distribution is characteristic of an early epidemic with fast cases predominating and is due to HIV infections, increased poverty, dismantling of control programs, and increased immigration from countries where TB is present. This young epidemic—due to recent transmission—now is superimposed on the mature one. This model of TB transmission makes two predictions: an increased population of newly developed cases should occur in younger individuals, and the proportion of primary infections (due to recent transmission) should also increase. Surveillance data validate these predictions. Importantly, the model also provides us with a cautionary note: if control strategies focus on preventing only new infections (and if reactivation or slow cases are neglected), it may take many decades to eliminate TB.

Controlling Consumption

Some believed TB to be an act of God, afflicting both rich and poor, against which there was no defense. More than 1,000 years ago, Hippocrates recommended a change in climate for phthisis, and this notion of the benefits of "clean air and sunshine" lasted well into the 20th century. Convinced that their infection was a result of bad air present in the crowded and dirty cities, many consumptives sought refuge in warmer or milder climates. In the United States there was a move to the Sunbelt of the West (Arizona, southern California, New Mexico, and Texas), while in Europe there was a move to the Mediterranean or South Africa. Ocean voyages were also recommended for those with TB because of the slow tempo and the clean air, especially on cruise ships. Living at high altitudes where the air was clear and crisp was considered beneficial, and so health seekers founded communities in Colorado. Some recovered, but others did not; a change of scene did not cure one's TB. Yet, in spite of the clear evidence that TB was not an environmental disease that could be cured by sunshine or fresh air, this view remained pervasive.

Prior to 1940 and as long as tuberculosis was diagnosed in its late phase, relief of symptoms became the prime method of treatment. There were many fads, and many of these did no good. Early treatments for TB included creosote, carbolic acid, gold, iodoform, arsenic, and menthol oil, administered either orally or as a nasal spray. Some physicians prescribed enemas of sulfur gases and drinking of papaya juice. None worked to cure TB. During the late 19th century, surgical techniques were used to "rest" the lung; this included pneumothorax or collapsed-lung treatments that required removal of several ribs and reduction in the size of the thoracic cavity. In some cases lung resections, i.e., removal of infected lung tissue, were performed. Despite a lack of success in curing the disease, these measures were continued well into the 1940s.

In the 1800s a movement developed both in Europe and in North America to treat TB patients in open-air hospitals—sanitaria (from "sana," meaning cure in Latin). Life in the sanitarium was a reaction to the stuffy, overheated rooms of the patient's home and workplace. Patients were made to take outdoor treatment in all weathers. Wide-open windows and outside balconies were the places where patients took the cure, and this was done in the hot summer as well as the freezing winter. The art of wrapping oneself in a blanket became an essential ritual. Fresh air was taken with a vengeance. Sanitaria were considered to be indispensable for cure, and patients were provided with brilliant sunshine, fresh air, quiet, rest, and good nutrition but no anti-TB medicines since none were available. Life in a luxury sanitarium (in Davos, Switzerland) is immortalized in Thomas Mann's novel *The Magic Mountain*, and one of the most famous of sanitaria in the United States was the Trudeau Institute in the Adirondack Mountains at Saranac Lake, N.Y., established by Edward Livingstone Trudeau. In 1873 Trudeau came down with TB after taking care of his consumptive brother, who had died 8 years earlier. Trudeau suffered from fever and weakness and began to cough up blood. Believing the end to be near, he decided to retire to the Adirondack Mountains, where he hiked, hunted, fished, swam, and anticipated death. However, within a few months of this outdoor regimen, he gained weight and recovered his energy, and his fever abated. In 1882, after Trudeau read an account of the benefits of a sanitarium in Europe, he raised money through donations and used his own funds to establish the first sanitarium in the United States at Saranac Lake. The sole industry of the town of Saranac Lake became the sanitarium.

The sanitarium movement soon went into high gear. In 1900 there were 34 sanitaria with 445 beds in the United States; 25 years later there were 356 sanitaria with 73,338 beds. However, sanitaria did not cure TB. In the prechemotherapy era (1938 to 1945) for sanitarium patients with advanced TB, the death rate was 69%, whereas for those with minimal disease it was 13%—about the same as that in the general population. The significant aspect of sanitaria is that they isolated contagious individuals and physicians maintained complete control of their patients.

Soon after Koch's discovery of the tubercle bacillus, antituberculosis campaigns began, first in Europe and then in North America. In 1889 to 1890, Hermann Biggs of the New York Department of Public Health issued an education leaflet that contained information on how to prevent the spread of consumption. Included were the following:

1. A campaign of education using newspapers and circulars calling attention to the dangers of TB, possibilities for treatment, and precautions to prevent it

2. Compulsory reporting of TB by all public institutions

3. Assignment of inspectors to visit the homes of patients in order to enforce sanitary regulations regarding the disposal of sputum and to arrange for disinfection

4. Provision of separate wards in hospitals for TB patients with lung disease

5. Provision of special consumptive hospitals to be used exclusively for the treatment of TB

6. Provision of laboratory facilities for the bacterial examination of sputum

Biggs' Riverside Hospital was established to confine, voluntarily or not, tubercular individuals whose "dissipated and vicious habits" endangered the health of the community. It was more a prison than a hospital, and in many respects this approach to TB was reminiscent of the establishment of leper colonies and hospitals for those suffering from leprosy. Biggs' idea was that protection of the public health was more important than individual freedom. He was the chief of the health police. Much of this attitude toward public health was based on a misguided meaning of Darwinian evolution: TB was no longer a romanticized disease, it was a sign of corruption, and pruning out the unfit would benefit mankind. Public health authorities impulsively attributed illness to the environment, i.e., how and where you lived, which was under your control. If you chose to live in filth, lacked ambition, or were too lazy to work yourself out of the tenements or were indolent and dozed by the fire instead of taking brisk walks in the fresh air and sunshine, then you brought TB on yourself. At all costs, the infected should be prevented from infecting the others in society. The TB that once had been romanticized and regarded as beautiful and a source of creative inspiration came to be regarded as a hereditary weakness. Succumbing to TB was, some contended, in a person's genes.

The United States declared a "War on Consumption" beginning in the early 1900s. This consisted of vigorous anti-TB campaigns, the most famous being Christmas Seals of the National Tuberculosis Foundation, which later became the American Lung Association. Early TB campaigns blamed capitalism for the disease because it was found in the crowded cities full of factory workers. At other times, TB victims were stigmatized by their "choice" of living in poverty; people claimed that those with TB were lazy, unambitious, and inferior. It is ironic that the whole TB eradication movement rose, flourished, and vanished without ever establishing its effectiveness against the disease. The greatest benefit of the War on Consumption, which was in full swing by 1915, was education, loss of stigma, abandonment of TB as a spiritualizing and romantic force, a greater understanding of contagion, the initiation of attempts to clean up tenements, and improvements in medical diagnosis and care.

Antibiotics to the Rescue

Prior to the 1940s, there were no drugs to cure TB. In 1939 Selman Waksman, a soil microbiologist and a Russian immigrant to the United States who was working at the Rutgers Agricultural College in New Jersey, began a systematic effort to identify soil microbes that could produce substances that might be useful in the control of infectious disease. Waksman based his program of antibiotic research on the earlier (1928) observations of Alexander Fleming. By 1940 Waksman and his graduate students had developed methods for growing soil microbes and screening them for their antibiotic properties. Waksman's graduate student Albert Schatz soon discovered that a mold named *Streptomyces griseus* (previously identified in soil samples by Waksman), obtained from the throat of a sick chicken that had been eating soil, was an antagonist and limited the survival of the TB bacteria both in soil and in sewage. Schatz developed methods to grow large amounts of *Streptomyces*, and in 1943 the inhibitory substance streptomycin was isolated from the cultures. Schatz's work was a part of his doctoral dissertation, and he and Waksman published their findings in 1944. Streptomycin inhibited the growth of tubercle bacilli both in the body and in test tubes. By 1945, it was in clinical use for treatment of TB, and by 1947 it was available in large quantities. In 1952, Waksman (but not Schatz) received the Nobel Prize for the discovery. Later, the mode of action of streptomycin was discovered: it inhibits the synthesis of the waxy cell wall of the bacillus, thus leaving the tubercle bacilli naked and unprotected from the onslaught of the killing machinery of the macrophage.

Unfortunately, the antibiotic-based victory over TB was short-lived. Soon there were signs of streptomycin-resistant TB bacilli. In 1949 streptomycin treatment was supplemented with *para*-amino salicylic acid (PAS), and in 1952 isoniazid, a drug first synthesized from coal tar in 1912, became the mainstay in the treatment of drug-resistant TB. Isoniazid also blocks the synthesis of the mycolic acids, which are a main constituent of the waxy wall of *M. tuberculosis*. Streptomycin and other drugs did not eradicate TB; however, they did effectively eliminate sanitaria. By 1954 the Trudeau Institute had closed, and by the 1960s almost all sanitaria had vanished.

In 1963, rifampin (derived from another *Streptomyces* species, *S. mediterranei*), an inhibitor of the synthesis of tubercle bacillus RNA, was introduced as a treatment. To minimize the emergence of drug resistance, patients are treated with a drug cocktail, called MDT (multiple-drug therapy). Individuals with latent TB are treated with isoniazid for 6 months, whereas patients who are also HIV infected require 9 months of treatment. Alternatively, treatment can be reduced to 4 months when isoniazid is combined with rifampin; however, patients coinfected with HIV require 2 months of treatment with rifampin and pyrazinamide. For patients with active TB, the treatment regimens involve isoniazid and rifampin for 9 months; if rifampin is not used, 18 months is the minimum duration of therapy for cure. Currently, the most commonly used treatment regimen is isoniazid, rifampin, and pyrazinamide administered daily for 8 weeks followed by isoniazid and rifampin given once, twice, or three times a week for 16 weeks. More than 85% of patients who receive both isoniazid and rifampin have negative sputum cultures within 2 months after treatment has begun. Isoniazid treatment for 6 to 9 months reduces TB prevalence by about 60% in HIV-infected individuals with a positive tuberculin test and about 40% when used irrespective of the skin test results. Regrettably, this low-cost intervention has been little used in Africa.

More than 50 years after the first effective use of chemotherapy for TB, the disease remains unconquered. The emergence of multidrug-resistant strains of *M. tuberculosis* has raised great concern because the disease caused by such strains is often fatal. In parts of South Africa, multidrug-resistant TB occurs in more than 2% of patients; in the United States, 0.9% of primary TB is multidrug resistant. These multidrug-resistant strains often develop in people who begin treatment but lapse after a few weeks, allowing larger numbers of mutant bacteria to survive; these mutants overcome and resist the drugs. In the late 1980s, it became clear that a substantial number of patients with TB were not completing treatment. To address this problem, the Centers for Disease Control and Prevention recommended that direct observation of therapy (DOT) by a trained health care worker be considered for all patients to ensure that they swallow each pill. Maintaining DOT is not easy, however, since it requires trained medical staff, infrastructure, and money. In practice, one-quarter of TB cases are not diagnosed before they become infectious and many who begin treatment do not stay the course.

There is evidence from twin studies that susceptibility to TB has a genetic basis, but no major TB susceptibility gene has ever been identified.

Susceptibility to TB is dramatically enhanced in HIV-infected persons, and the spread of AIDS has been paralleled by a resurgence in TB cases. HIV-infected individuals who adhere to the standard regimen for TB do not have an increased risk of treatment failure or relapse. However, the use of protease inhibitors and reverse transcriptase inhibitors for HIV treatment has complicated the treatment of TB in HIV-infected individuals. Administration of these drugs with rifampin can result in lower levels of antiviral drugs and toxic levels of rifampin. Rifabutin may be substituted for rifampin with fewer side effects in combination with some antiviral drugs but not with others.

A Vaccine for TB

Why is there no vaccine against TB? There is. In the 1920s two French bacteriologists, Albert Calmette (1863 to 1933) and Camille Guérin (1872 to 1961), used a technique first used in 1882 by Louis Pasteur—they attenuated bovine TB (*M. bovis*). They grew the bacteria on a nutritive medium containing beef bile for 231 generations over a period of 13 years, during which time the strain became weakened. This non-disease-producing bacillus, which could elicit immunity, became the vaccine called BCG (bacillus Calmette-Guérin) and was cross-reactive with human TB. When the vaccine was first introduced in a trial in Lübeck, Germany, in 1930, 249 babies were inoculated and 76 died. This was not due to inoculation with BCG itself but to contamination with live, virulent *M. tuberculosis* stored in the same incubator; however, the adverse publicity so poisoned the public's interest that it was difficult to gain popular support for BCG vaccination for decades. This accident, however, may have had one beneficial effect: it encouraged the search for anti-TB drugs.

Injection of BCG produces a mild infection, induces immunity, and has never resulted in a virulent infection. It has successfully reduced TB-related mortality by about 90% in vaccinated children; however, it has had little effect on pulmonary TB, which is most common in young adults in regions where tuberculosis is endemic. Furthermore, attempts to extend the period of protection by giving a booster dose of the vaccine have been unsuccessful. The reasons for this remain unclear but may be due to a low level of immunity induced by exposure to soil mycobacteria or may be because previous vaccination with BCG is sufficient to inhibit the growth of BCG, blocking its boosting effect, but has only a small effect on the more virulent strains of *M. tuberculosis*. Although the mortality from TB in countries that have used BCG since 1950 has declined, there has also been a marked decline in countries not using BCG. In the United States, many are reluctant to use a live vaccine; also, because vaccination with BCG renders the tuberculin test positive, this diagnostic tool cannot be used. Furthermore, the use of BCG as a vaccine would be of little benefit to those already infected.

Consequences

Tuberculosis is the leading cause of death from a curable infectious disease. In many resource-poor countries, especially those blighted with HIV, TB is on the rise, with one person infected every second. Although the death rate is low (<0.01%), it is expected that 30 million people will die of TB in the next decade. However, TB is largely a forgotten disease in the United States, having been replaced by cancer, AIDS, and cardiovascular diseases in the public awareness. TB still kills, and the new strains are more dangerous because they are drug resistant.

It is estimated that one-third of people worldwide are presently infected. In 2004 there were 89 million new cases and 1.7 million deaths, with all but 400,000 occurring in developing countries. About 3.9 million cases were sputum-smear positive cases—the most infectious type. The majority of patients with TB live in the most populous countries of Asia; Bangladesh, China, India, Indonesia, and Pakistan account for half of the new cases.

Globally, four epidemiological patterns can be discerned. In developing countries with low rates of HIV, TB infection responds well to control programs. In the industrialized world TB is increasingly attributed to immigration, whereas high rates of multidrug-resistant TB, economic decline, and substandard health services threaten measures for control in parts of Eastern Europe. The dominant challenge remains in sub-Saharan Africa. In Africa, TB is the first manifestation of HIV infection and is the leading cause of death in HIV-infected patients. HIV prevalence in TB patients is 38% in Africa; in countries with the highest rates of HIV prevalence, more than 75% of cases of TB are HIV associated. TB incidence increases with worsening immunosuppression, so that rates are 8.3 times higher in HIV-positive than HIV-negative Africans. In six southern African countries with adult HIV prevalence of more than 20%, the case rates for TB are 460 to 720 per 100,000 per year, whereas in the United States the rate is 5 per 100,000 per year. Ironically, HIV in Africa might contribute much less to disease transmission because of early diagnosis and death, yet it contributes greatly to the incidence of TB and death.

Worldwide, TB control has been based on the World Health Organization-promoted DOT strategy, which involves prompt diagnosis and effective treatment of individuals who have smear-positive TB in order to reduce the rates of transmission. However, in Africa this strategy may not be effective because even a rigorous program cannot adequately compensate for the rising susceptibility to TB as HIV rates soar. Further complicating control is the fact that HIV-related TB is more usually smear negative, extrapulmonary, or disseminated. Better control could be accomplished if higher priorities were given to smear-negative disease and starting anti-TB treatment earlier and with appropriate follow-up and outcome assessment.

In the United States since 1984 there has been a large increase in TB cases in men aged 25 to 44, the group at greatest risk for AIDS. In the United States the rate of TB reached a peak in 1992 and has decreased steadily since then. In 2005 there were 14,093 cases in the United States. The CDC statistics indicate that 62% of the TB cases are in African-Americans, Hispanics, and Asians. In these cases, 40% were under age 35, whereas in Caucasians it is primarily a disease of the elderly, especially those in nursing homes. Over 20% of the U.S. cases occur in foreigners who come from resource-poor countries.

What measures can be instituted to effectively control TB? Although DOT has contained and reduced multidrug-resistant TB in some places in the world, the standard treatment regimens and direct observation may not be sufficient to contain the disease in Africa. In Africa, more comprehensive surveillance, better assessment of drug sensitivity, implementation of fixed drug combination tablets, increased numbers of health care workers, and establishment of policies to manage multidrug-resistant TB in the HIV setting will be needed. "It is imperative that there not be a collapse of previously well-functioning TB control programs after competition of scarce human resources and to ensure the integration of HIV and TB services does not compromise core TB program functions, such as maintenance of drugs and supplies, prevention of drug resistance, assurance of quality diagnostic microscopy (of sputum) and cohort analysis of treated patients."

When humans were hunter-gatherers and lived in small roving bands, TB was not a serious threat; however, with the development of agriculture and animal husbandry, the numbers of people increased, as did exposure to new pathogens. A soil bacterium infected cattle or game and it became *M. bovis*, and perhaps by crossing the species barrier it evolved into *M. tuberculosis*. Coincident with urbanization were epidemics of TB. The poor—undernourished, crowded, and living under unhygienic conditions—became the breeding ground for the "white plague." Today it is recognized that tuberculosis is an infectious and a societal disease. Understanding this contagious illness demands that social and economic factors be considered as much as the way in which the tubercle bacillus causes damage to the human body, how it manages to evade the immune system, and how it is able to overcome the most potent anti-TB drugs. In the past, fear of the disease, fueled by preexisting prejudices, led to public and institutional reactions including mandatory testing and isolation and tended to stigmatize immigrants and those with different lifestyles. Tuberculosis has affected the emotional and intellectual climate of human populations throughout the world and will continue to do so. Now, almost 125 years after the discovery of the cause of TB, it must be realized that for "The Peoples Plague" to be contained, the subtle interplay between disease and society must be fully appreciated. Until that time comes, TB will remain a disease that will threaten us.

8

Malaria

I am Ronald Ross a physician in the Indian Medical Service currently working in Secunderbad, India. On August 16th, my assistant brought me a bottle that contained about a dozen big brown mosquitoes, with fine tapered bodies hungrily trying to escape through the gauze covering of the flask. My mind was blank with the August heat; the screws of the microscope were rusted with sweat from my forehead and hands, while the last remaining eyepiece was cracked. I fed them on a patient with malaria. There were some casualties among the mosquitoes, with only a few left on the morning of August 20th. At about 1 p.m., I determined to sacrifice the last mosquito. Was it worth bothering about the last one, I asked? And, I answered myself, better finish off the batch. A job worth doing at all is worth doing well. The dissection was excellent and I went carefully through the tissues, now so familiar to me, searching every micron with the same passion and care as one would have in searching some vast ruined palace for a little hidden treasure. Nothing. No, these new mosquitoes also were going to be a failure. But the stomach tissues still remained to be examined—lying there, empty and flaccid, before me on the glass slide, a great white expanse of cells like a large courtyard of flagstones, each one of which must be scrutinized. I was tired and what was the use? I must have examined the stomachs of a thousand mosquitoes by this time. I had scarcely commenced the search again when I saw a clear and almost perfectly circular outline before me. The outline was too sharp, the cell too small to be an ordinary stomach cell of a mosquito. I looked a little further. Here was another, and another exactly similar cell. In each of these, there was a cluster of small granules, black as jet. I made little pen-and-ink drawings of the cells with black dots. The next day, I wrote the following verses:

This day relenting God
Hath placed within my hand
A wondrous thing; and God
be praised. At his command,

Seeking his secret deeds
With tears and toiling breath,
I find thy cunning seeds,
O million-murdering death.

I know this little thing
A myriad men will save.
O death, where is thy sting?
Thy victory, O grave?

Here was the clue . . . four or five days after feeding on malaria-infected blood, the mosquito itself had become infected—there were pigmented swellings (oocysts) embedded in the stomach wall. But, did these keep on growing, and how did these mosquitoes become infective to humans?

I knew I could complete the work in a few weeks but before I could accomplish this I was ordered to Calcutta to deal with a cholera outbreak. In Calcutta I was provided with a laboratory and it was my intention to continue work on human malaria, however, since there were not a large number of human cases in the hospitals I was forced to study malaria in birds. Pigeons, crows, larks and sparrows were caught and placed in cages on two old hospital beds. Mosquito nets were put over the beds and then, at night, infected mosquitoes were put under the nets. Before much time had passed the crows and pigeons were found to have malaria parasites in their blood.

One day, while studying some sparrows I found one was quite healthy, another contained a few of the malaria parasites, and the third had a large number of parasites in its blood. Each bird was put under a separate mosquito net and exposed to a group of mosquitoes from a batch that had been hatched out from grubs. Fifteen mosquitoes were fed on the healthy sparrow; in their stomachs, not one parasite was found. Nineteen mosquitoes were fed on the second sparrow; every one of these contained some parasites, though, in some cases, not very many. Twenty insects were fed on the third, badly infected, sparrow; every one of these contained some parasites in their stomachs and some contained huge numbers.

This delighted me! But I still did not have the full details of . . . the change from the oocysts in the mosquito's stomach into the stages that could infect human beings and birds. Then, I found that some of the oocysts seemed to have stripes or ridges in them; this happened on the 7th or 8th day after the mosquito had been fed on infected blood. I spent hours every day peering into the microscope. The constant strain on mind and eye at this temperature is making me thoroughly ill. I thought no doubt these oocysts with the stripes or rods burst—but then what happened to them? Then, on July 4th, I got something of value. Near a mosquito's head there was a large branch-looking gland. It led into the head of the mosquito. It is a thousand to one that it is a salivary gland. Did this gland infect healthy creatures? Did it mean that if an infected mosquito fed off the blood of an uninfected

> human being or bird, then this gland would pour some of the parasites (called sporozoites) . . . into the blood of the healthy creature?
>
> During July 21 and 22, I took some uninfected sparrows, allowed mosquitoes (which had been fed on malaria-infected sparrows) to bite them and then, within a few days, was able to show that the healthy sparrows had become infected. This was the proof—this showed that malaria was not conveyed by dust or bad air. On July 28, 1898 I reported how mosquitoes transmit malaria. This finding, after 17 years of painstaking effort, would lay the foundation for the application of practical methods to control the disease. And, for the first time in history it was possible to show that if the mosquito vector could be eliminated, malaria transmission could be prevented.

Ross would be unhappy to learn that malaria has not yet been eliminated. Today, every 10 s a person dies of malaria—mostly children under the age of 5 years living in Africa. The total number of cases of malaria is estimated to be 300 million to 500 million, with approximately 10% of these occurring outside Africa. Annually two million to three million deaths are caused by malaria. Malaria infections are on the rise in many parts of the world, and the disease continues to affect us in the places we live, work, travel to, and fight in.

Origins

Malaria is an ancient disease. Records in the Ebers Papyrus (ca. 1550 BC), in clay tablets from the library of Ashurbanipal (ca. 668–627 BC), and in the classic Chinese medical text the Nei Ching (AD 100) describe the typically enlarged spleen, periodic fevers, headache, chills, and fever. Malaria probably came to Europe from Africa via the Nile Valley or because of closer contact between Europeans and the people of Asia Minor. The Greek physician Hippocrates (460 to 370 BC) discussed, in his *Book of Epidemics*, the two kinds of malaria: one with recurrent fevers every third day (benign tertian) and another with fevers every fourth day (quartan). He also noted that people living near marshes had enlarged spleens. Although Hippocrates did not describe malignant tertian malaria in Greece, there is clear evidence of the presence of this malaria in the Roman Republic by 200 BC. Indeed, the disease was so prevalent in the marshland of the Roman Campagna that the condition was called the "Roman fever." Since it was believed that this fever recurred during the sickly summer season due to vapors emanating from the marshes, it was called by the Italian name mal' aria, literally, "bad air." Over the centuries malaria spread across Europe, reaching Spain and Russia by the 12th century; by the 14th century it was in England. Malaria was brought to the New World by European explorers, conquistadors, colonists, and African slaves. By the 1800s it was found worldwide.

"Bad air" was considered the cause of malaria until 1880. On 20 October of that year, Alphonse Laveran (1845 to 1922), a military physician stationed in Bone, Algeria, examined a drop of blood from a soldier suffering from an intermittent fever. Under the light microscope Laveran noticed among the red blood cells some crescent-shaped bodies that were almost entirely transparent save for some pigment inclusions. On 6 November 1880, while examining a drop of blood from a feverish artilleryman, he saw several transparent mobile filaments emerging from a clear spherical body. He recognized that these bodies were alive and that he was looking at an animal parasite, not a bacterium or a fungus. Subsequently he examined blood samples from 192 malaria patients: in 148 of these samples he found the telltale crescents. Where there were no crescents, there were no symptoms of malaria. He named the parasite *Oscillaria malariae* and communicated his findings to the Societe Medicale des Hospitaux on 24 December 1880. The drawings in his paper provide convincing evidence that, without use of stains or a microscope fitted with an oil immersion lens, Laveran had seen the development of the parasite. At first, Laveran's announcement was received with skepticism. Indeed, in 1882 when he visited Rome and showed his slides to the Italian parasitologists, they scoffed and told him that the spherical bodies were nothing more than degenerating red blood cells. Initially, the Italians examined only preparations that had been heat fixed and stained with methylene blue, so they did not see the movement that had caused Laveran to name the parasite *Oscillaria*. However, 2 years later, when, like Laveran, they began to examine fresh preparations of blood, they were able to observe the ameboid movements of the parasite within red blood cells, as well as emerging whiplike filaments from the clear spherical bodies within the red blood cells. In 1886, using thin smears of fresh blood, Camillo Golgi (1843 to 1926) discovered that the parasite reproduced asexually by multiple fission and showed that the fever coincided with erythrocyte lysis and parasite release. In 1891, Dimitri Romanowsky prepared heat-fixed thin blood films and used a combination of methylene blue and eosin to stain the parasite. The significance of Laveran's observation of the release of motile filaments went unappreciated until 1896 to 1897, when William MacCallum and Eugene Opie, students at Johns Hopkins University, found that the blood of sparrows and crows infected with *Haemoproteus* (a bird parasite closely related to the agent of malaria) contained two kinds of crescent-shaped gametocytes (sex cells) and that filament extrusion (called exflagellation) reflected the release of microgametes from the male gametocyte. They also correctly interpreted their observations: gametocytes occur in the blood, and when ingested by a biting fly, the gametes are released in the stomach, where fertilization takes place, producing a wormlike zygote, the ookinete. Neither Laveran nor MacCallum solved the problem of malaria transmission; that was left for Ronald Ross.

The Disease

Although some 170 kinds of malaria have been described, only 4 are specific for humans. The human malaria parasites, *Plasmodium falciparum, P. vivax, P. ovale,* and *P. malariae,* are transmitted through the bite of an infected female anopheline mosquito when, during blood feeding, she injects sporozoites from her salivary glands. Usually, fewer than 50 sporozoites are inoculated. These travel via the bloodstream to the liver, where they enter liver cells. The entire process takes less than 1 h. Within the liver cells, each parasite multiplies asexually to produce 10,000 or more infective offspring. These do not return to their spawning ground, the liver, but instead invade erythrocytes. It is the asexual reproduction of parasites in red blood cells and their ultimate destruction with the release of infectious offspring (merozoites) that is responsible for the pathogenesis of this disease. Merozoites released from erythrocytes can invade other red cells and continue the cycle of 10-fold parasite multiplication, with extensive red blood cell destruction. In some cases the merozoites enter red cells but do not divide. Instead, they differentiate into male or female gametocytes (the crescents of Laveran). When ingested by the female mosquito, the male gametocyte divides into eight flagellated microgametes, which escape from the enclosing red cell (in a process called exflagellation); these microgametes swim to the macrogamete, one fertilizes it, and the resultant motile zygote, the ookinete, moves either between or through the cells of the stomach wall. This encysted zygote, resembling a wart on the outside of the mosquito stomach, is an oocyst, and through asexual multiplication threadlike sporozoites are produced therein. The oocyst bursts, releasing its sporozoites into the body cavity of the mosquito, and the sporozoites quickly find their way to the salivary glands. When this female mosquito feeds again, the transmission cycle is completed.

All of the pathology of malaria is due to parasite multiplication in erythrocytes. Here is what it feels like:

> "I wanted to sit up, but felt that I didn't have the strength to, that I was paralyzed. The first signal of an imminent attack is a feeling of anxiety, which comes on suddenly and for no clear reason. Something has happened to you, something bad. If you believe in spirits, you know what it is: someone has pronounced a curse, and an evil spirit has entered you, disabling you and rooting you to the ground. Hence the dullness, the weakness, and the heaviness that comes over you. Everything is irritating. First and foremost, the light; you hate the light. And others are irritating—their loud voices, their revolting smell, their rough touch. But you don't have a lot of time for these repugnances and loathings. For the attack arrives quickly, sometimes quite abruptly, with few preliminaries. It is a sudden, violent onset of cold. A polar, arctic cold. Someone has taken you, naked, toasted in the hellish heat of the Sahel and the Sahara and has thrown you straight into the icy highlands of Greenland or Spitsbergen, amid the snows, winds, and blizzards. What a shock! You feel the cold in a split second, a terrifying, piercing, ghastly cold. You begin to tremble, to quake to thrash about. You immediately recognize, however, that this is not a trembling you are familiar with from earlier experiences—when you caught cold one winter in a frost; these tremors and convulsions tossing you around are of a kind that any moment now will tear you to shreds. Trying to save yourself, you begin to beg for help. What can bring relief? The only thing that really helps is if someone covers you. But not simply throws a blanket or quilt over you. The thing you are being covered with must crush you with its weight, squeeze you, flatten you. You dream of being pulverized. You desperately long for a steamroller to pass over you. A man right after a strong attack . . . is a human rag. He lies in a puddle of sweat, he is still feverish, and he can move neither hand nor foot. Everything hurts; he is dizzy and nauseous. He is exhausted, weak, and limp. Carried by someone else, he gives the impression of having no bones and muscles. And many days must pass before he can get up on his feet again."

The fever spike may reach 41°C and corresponds to the rupture of the red cells as merozoites are released from them. Anemia is the most immediate pathologic consequence of parasite multiplication and destruction of erythrocytes, and there can also be suppression of red cell production in the bone marrow. During the first few weeks of infection, the spleen is palpable because it is swollen from the accumulation of parasitized red cells. At this time it is soft and easily ruptured. If the infection is treated, the spleen returns to normal size; however, in chronic infections the spleen continues to enlarge, becoming hard and blackened due to the accumulation of malaria pigment. The long-term consequences of malaria infections are an enlarged spleen and liver as well as organ dysfunction.

Falciparum infections are more severe and, when untreated, can result in a mortality rate of 25% in adults. In pregnant females, falciparum malaria may result in stillbirth, lower than normal birth weight, or spontaneous abortion. Nonimmune persons and children may develop cerebral malaria, a consequence of the mechanical blockage of microvessels in the brain due to sequestration of infected red cells. If relapse occurs, it is due to the increase in the numbers of preexisting erythrocytic forms, previously too small to be detected microscopically. Falciparum malaria accounts for 50% of all clinical malaria cases and is responsible for 95% of malaria-related deaths. *P. vivax* and *P. ovale* malarias also have the capacity to relapse; that is, parasites can reappear in the blood after a period when none were present. This type of relapse is due to the delayed liberation of merozoites from preerythrocytic stages in the liver, called hypnozoites. *P. vivax* results in severe and debilitating attacks but is rarely fatal. It accounts for about 45% of all malaria cases. The benign quartan malaria, due to *P. malariae*, may persist in the body for up to four decades without causing signs of pathology. *P. malariae* and *P. ovale* infections are responsible for approximately 5% of all clinical cases.

"Catching" Malaria

Ross' discovery of infectious stages in the mosquito salivary glands in a bird malaria appeared to be the critical element in understanding the transmission of the disease in humans. However, Ross' mentor Sir Patrick Manson rightly cautioned: "One can object that the facts determined for birds do not hold, necessarily, for humans." Ross and Manson wanted to grab the glory of discovery for themselves and for England, but they were not alone in such a quest. The German government dispatched a team of scientists under the leadership of Robert Koch to work in the Roman Campagna, an area notorious for endemic malaria. They isolated a bacterium from the air and the mud of the marshes and, rejecting the claims of Laveran, named the causative agent for malaria *Bacillus malariae*. However, when the bacillus could not be grown in the laboratory, Koch discarded it as the cause of the disease; undeterred, he continued to look further. He then visited the laboratory of Giovanni Battista Grassi at the University of Rome, told him of his failure with the "germ" of malaria, and mentioned

Ross' communication. At that moment Grassi had what is today called an "aha" moment. Where Ross was patient and willing to carry out a seemingly endless series of trial-and-error experiments, Grassi was methodical and analytical—he was also able to distinguish the different kinds of mosquitoes. Grassi observed that "there was not a single place where there is malaria—where there aren't mosquitoes too, and either malaria is carried by one particular blood sucking mosquito out of the forty different kinds of mosquitoes in Italy—or it isn't carried by mosquitoes at all." He recognized that there were still two tasks left: identify the mosquito that transmitted human malaria and then demonstrate the mosquito cycle for human malaria. Working with Amico Bignami, Giovanni Bastianelli, Angelo Celli, and Antonio Dionisi, he went into the highly malarious Roman Campagna and the surrounding area, collecting mosquitoes and at the same time recording information on the incidence of malaria among the people. (Grassi was, in effect, carrying out an epidemiologic study.) It soon became apparent that most of the mosquitoes could be eliminated as carriers of the disease because they were found where there was no malaria. However, there was an exception. Where there were "zanzarone," as the Italians called the large brown spotted-wing mosquitoes, there was always malaria. Grassi recognized that the zanzarone were *Anopheles*, and he wrote, "It is the anopheles mosquito that carries malaria . . ." As Grassi correctly observed, "Mosquitoes without malaria . . . but never malaria without mosquitoes." Grassi and his team were able to infect clean anopheles mosquitoes by having them feed on patients with crescents in their blood, and they could trace the development of the parasite from the mosquito stomach to the salivary glands. The life cycle in the human was, as Ross had correctly surmised, similar to that of the bird malaria with which he had worked. Grassi's work showed that the association of the disease with swampy, marshy areas of the world is due to the fact that these areas are ideal breeding sites for mosquitoes.

With this work, Grassi demolished the theory of Koch and was able to prove that "It is not the mosquito's children, but only the mosquito who herself bites a malaria sufferer—it is only that mosquito who can give malaria to healthy people." Grassi wanted his beloved country, Italy, to receive recognition for his work, but it did not. Instead, it provoked a bitter and very nasty disagreement with Ross. Ross claimed that it was only after Grassi had read his work on the transmission of malaria using birds that he recognized that only in areas containing *Anopheles* was there human malaria, and that *Culex* was not involved, but Grassi did not publish this or the development of the parasite in these mosquitoes until late in 1898. Ross wrote in his Memoirs, "They . . . had this paper of mine before them when they wrote their note. Their statement was . . . a deliberate and intentional lie, told in order to discredit my work and so to obtain priority . . . Many of the items . . . are directly pirated from my . . . results . . . stolen straight from me . . ." Ross did not complete the proof of mosquito transmission with human malaria; the Italians did that. However, he flagged the dapple-winged mosquito as the vector. Grassi's contribution was to recognize the vector as *Anopheles*. In effect, Ross was the explorer at the helm of the ship and the Italians rode the decks and helped make a landing. Ross, not Grassi, received the Nobel Prize in 1902, and Laveran received the Prize in 1907. The embittered and unforgiving Ross died in 1932.

Control

Once Ross and Grassi had flagged the *Anopheles* mosquito as the vector of malaria, interest turned toward methods for control of both the disease and the vector. Of the 450 species of *Anopheles*, only 50 are capable of transmitting the disease; of these, only 30 are considered efficient vectors. In Africa the most efficient vector is *Anopheles gambiae*, which can breed in small temporary pools of water such as those formed by foot or hoof prints or tire tracks. In other areas, the vector is *A. stephensi*, which can breed in wells or cisterns.

Generally speaking, to prevent malaria, infected mosquitoes must be blocked from feeding on humans, the breeding sites of mosquitoes must be eliminated, and measures to kill mosquito larvae, as well as to reduce the life span of the blood-feeding adult, must be implemented. Contact with adult mosquitoes can be prevented by using insect repellents, wearing protective clothing, using impregnated mosquito netting, and screening houses. Breeding sites can be controlled by draining water, changing its salinity, flushing, altering water levels, and clearing vegetation. Adult mosquitoes can be killed by using insect sprays, and larvae can be destroyed with larvicides. In the 1900s, larvicides in the form of oil and Paris green (bright green powdered copper acetoarsenite, which is extremely poisonous and is sometimes used as an insecticide or fungicide), along with drainage, were introduced to limit mosquito-breeding sites in water. This had outstanding success in reducing transmission in some parts of the world. Later, DDT was introduced as a component of an eradication campaign; however, by the early 1960s it became clear that eradication could not be accomplished due to the emergence of DDT-resistant mosquitoes and the negative ecological side effects of DDT. By 1969 the World Health Organization formally abandoned its eradication campaign and recommended that countries employ control strategies. Today, attempts at control involve using insecticide-impregnated bed nets and spraying with ecologically less disruptive but more expensive insecticides.

Chemotherapy

Since all of the pathology of malaria is due to parasites multiplying in the blood, most antimalarials are directed at these rapidly dividing stages. The earliest of the antimalarials was quinine, derived from the bark of the cinchona tree and isolated by two French chemists, Pierre Pelletier and Joseph Caventou, in 1817. Quinine continues to be used, but completion of the 5- to 7-day regimen for cure is poor due to unpleasant side effects such as bitter taste, tinnitus, nausea, and vomiting, and so it remains limited to parenteral use. Chloroquine and amodiaquine are synthetic antimalarials developed in the 1940s. They were the mainstay of the unsuccessful malaria eradication program of the 1950s. Both act rapidly and can be taken prophylactically, by mouth, once a week. However, some strains of *P. falciparum* have become resistant to chloroquine. Mefloquine (trade name Lariam) and halofantrine (trade name Halfan) were synthesized in the 1960s as alternatives to chloroquine; mefloquine, which acts similarly to quinine, can be taken orally once a week before, during, and after exposure. Primaquine is used against the stages in the liver in the vivax malarias, and if treatment is successful it prevents relapse. A newer analog of primaquine called tafenoquine has been developed by a collaboration between the U.S. Army and GlaxoSmithKline; it is the first new replacement drug since primaquine was introduced more than 60 years ago.

Regrettably, little is known about the molecular mode of action of many of these drugs or the mechanisms of resistance, although it has been contended that multidrug resistance to halofantrine, mefloquine, chloroquine, and quinine is due to mutations. We know quite clearly that some antimalarials block the synthesis of parasite DNA, similar to the action of AZT in HIV infections. In the late 1980s, Burroughs Wellcome began a program for the rational design of antimalarials. The result was atovaquone, a mimic of the vitamin coenzyme Q or ubiquinone; it also blocks DNA synthesis by the parasites. Atovaquone combined with proguanil to form the drug Malarone is used as a prophylactic antimalarial.

Qinghaosu, a Chinese herbal medicine, is both the newest and the oldest in the arsenal of antimalarials. It has been used in China for 2,000 years to reduce fever. It is derived from the leaves of the wormwood *Artemisia annua* and is also known as artemisinin. Artemisinin itself is poorly absorbed, so three derivatives are widely used outside the United States: artemether, artemotil, and artesunate. These act rapidly to clear the blood of parasites, but there is a high rate of relapse.

The major threat of malaria today is not an increasing range of endemicity but, rather, a rise in the intensity of antimalarial drug resistance. Drug resistance in malaria has been defined as the ability of a parasite strain to survive and/or multiply despite the administration and absorption of a drug in doses equal to or higher than those usually recommended but within the limits of tolerance of the subject. Thus, resistance is a characteristic of the particular parasite strain. First recognized more than 40 years ago with chloroquine in South America and Southeast Asia, drug-resistant malaria poses one of the greatest challenges for controlling morbidity and mortality; its treatment is dependent on prompt and accurate diagnosis. In an attempt to overcome or delay the emergence and spread of drug-resistant strains, combination therapy (with Fansidar, which combines sulfadoxine and pyrimethamine; Fansimef, which utilizes Fansidar plus mefloquine; Maloprim, which is pyrimethamine plus dapsone; and Malarone, which combines atovaquone and proguanil) has been deployed. Since 1991, GlaxoSmithKline has been developing LAPDAP, a combination of lapudrine or chlorproguanil with dapsone.

In addition to resistance, cost is another treatment constraint. Chloroquine was a very cheap antimalarial, costing about 8 cents per treatment, whereas a course of treatment with the newer drugs such as mefloquine may be 10 times as high, halofantrine may cost 20 to 30 times as much as chloroquine, and a course of treatment with atovaquone may cost up to $30. LAPDAP, released in 2002, costs less than $1 per tablet and may provide a cheaper alternative. A third constraint for the drug treatment of malaria is the reluctance of pharmaceutical companies to invest their capital in developing antimalarials that they believe will not yield profits.

The Elusive Malaria Vaccine

What makes scientists believe that a vaccine against malaria is both practical and possible? First, there is abundant evidence from natural infections of humans that immunity is acquired, and although it is incomplete (i.e., nonsterilizing) and ineffective in preventing reinfection, immunity does result in a reduction in mortality. Second, adults in areas of endemic infection produce antibody and have low fatality rates. Third, immunization with radiation-attenuated sporozoites induces sterile immunity and protects >90% of human recipients for more than 10 months. Fourth, passive transfer of immunoglobulins, purified from adults who are immune, protects the recipient against disease.

A malaria vaccine is yet to be developed; however, when it does become available, it will have to be safe and easy to administer. To be fully protective, the vaccine will have to produce a strong immune response. In parts of Africa where malaria is endemic and the value for R_0 is 50 to 100 (see p. 170–171), it would require 99% coverage using a lifelong vaccine given at 3 months of age to eliminate malaria. Making the vaccine situation even more difficult is that we do not have an in vitro correlate for protective immunity and the availability of suitable monkeys for testing a vaccine is not great.

Nevertheless, work goes on to develop a malaria vaccine. Some of the potential parasite targets for a vaccine are (i) liver stage vaccines that reduce the chance that a person will become sick (this would be suitable for travelers and the military); (ii) blood stage vaccines to reduce disease severity and the risk of death; and (iii) mosquito stage vaccines to prevent the spread of malaria through the community (these are also called transmission-blocking or altruistic vaccines). The vaccines being contemplated are not based on killed or attenuated stages, but, instead, attention is focused on subunit vaccines consisting of selected antigens. Such subunit vaccines may consist of synthetic peptides, recombinant proteins, or parasite DNA packaged in a virus or naked. The World Health Organization claims that there are over 100 malaria vaccines under development.

Consequences

One of the Four Horsemen of the Apocalypse, Pestilence, in the form of malaria has long been a companion of War. Alexander the Great, having conquered most of the known world, did not extend his conquests over the entire subcontinent of India in large part because in 323 BC he died of malaria at age 33. Malaria repelled foreign invaders from sacking ancient Rome, and Caesar's campaigns were disrupted by malaria. Frederick Barbarossa's army in the 12th century was also prevented from attacking Rome because it was felled by "bad air." Malaria (then called ague) was prevalent in the fens of England and in colonial America, debilitating the farming population. Indeed, in some instances it was the major obstacle both socially and economically in the growth and development of the American colonies. In the time of the Revolutionary War, malaria played a role in several critical battles, and during the Civil War there was a high incidence of malaria among the troops from the North stationed in the South, with an estimated one-half of the white troops and four-fifths of the black troops contracting malaria annually. Malaria plagued the French and British engaged in battle during World War I and again in World War II. In 1943 Sir William Slim, the British Field Marshal, said, "For every man evacuated with wounds we had one hundred and twenty evacuated sick." Malaria, the most significant disease problem facing the Allies, prompted one writer to claim "that the triumph over malaria was one of the most important victories of allied forces in the Southwest Pacific." Again, during the Korean and Vietnam wars, malaria was "The Great Debilitator," with *P. vivax* causing the greater morbidity in the former war and *P. falciparum* causing greater morbidity in the latter. This prompted the generals to provide better antimalarial measures and also stimulated the search for new antimalarials. The problems facing medical planners in future conflicts, however, will be even greater because the next time around, multidrug-resistant strains of *P. falciparum* and insecticide-resistant mosquitoes will have to be overcome. Furthermore, U.S. forces have only infrequently had to contend with malaria in sub-Saharan Africa, where the number of infected bites can be 100 times greater than other areas where we have fought.

Malaria is by far the most important of the world's tropical parasitic diseases, but it can (and does) also exist in temperate areas. At present 90 countries or territories in the world are considered malarious, with almost half in Africa south of the Sahara. Today, malaria still ravages the continent of Africa; its target is the indigenous peoples, and slowly and inexorably it is killing them. This is especially true in the rainy and low-lying agricultural areas, where transmission is high. In other areas the problem is famine and malnutrition. Malnutrition increases susceptibility to malaria and a variety of other diseases, leads to lower productivity, and puts a further strain on the already fragile health care system. Sadly, in Africa, the Horsemen of the Apocalypse have as their principal target the children; one in three children die of malaria, and a like number die of AIDS.

9

Yellow Fever: the Saffron Scourge

In 1926 Paul de Kruif, the author of *Microbe Hunters*, wrote,

> "... everybody knew just how to fight that most panic striking plague, yellow fever; everybody had a different idea on how to combat it. You should fumigate silks and satins and possessions of folks before they left yellow fever towns—no! that is not enough: you should burn them. You should bury, burn and destroy these silks and satins before they come into yellow fever towns. It was not wise to shake hands with friends whose families were dying of yellow fever; it was perfectly safe to shake hands with them. It was best to burn down houses where yellow fever lurked—no! it was enough to smoke them out with sulfur. But there was one thing everybody . . . agreed upon for nearly two hundred years, and that was this: when folks of a town began to turn yellow and hiccup and vomit black by scores, by hundreds, every day—the only thing to do was to get up and get out of town. This was the state of scientific knowledge about yellow fever up to the year 1900." And, he continued, "Everybody is agreed that Walter Reed—head of the Yellow Fever Commission . . . had to risk human lives; animals simply will not catch yellow fever! . . . there are no arguments—and that makes it fun to tell this story of yellow fever. It vindicates Pasteur! Because . . . there is hardly enough of the poison of yellow fever left in the world to put on the points of six pins; in a few years there may not be a single speck . . . left on earth—it will be as completely extinct as the dinosaurs. . . ."

Despite de Kruif's unbridled optimism as well as the success of Reed's "gruesome experiments" using human volunteers, the "poison" of yellow fever has not become extinct. Annually about 1,000 cases of yellow fever are reported worldwide, but the actual number may be 200 times greater, and with a reservoir of jungle yellow fever in monkey populations in tropical South America and Africa, the disease will never be eradicated. There is a further conundrum: although yellow fever has never appeared in Asia, what would be the consequences if it did?

The Disease

In *Guinea Pig Doctors. The Drama of Medical Research Through Self-Experimentation*, Jon Franklin and John Sutherland give a graphic description of the disease:

> "Yellow fever had a macabre way of toying with its victim before killing. . . . For three days there was fever and chills followed by a marked improvement. The temperature fell. . . . He could think, if it pleased him to fantasize, that the worst was past. Maybe he could tell himself it had been influenza. Or perhaps a touch of malaria. Or some unnamed tropical thing, of which there were many. But on the fourth day yellow fever returned with a vengeance. Beads of sweat popped out on the victim's skin, as the fever returned and climbed steadily to 103, 104, 105 . . . and then the chills came and the victim's teeth chattered and he begged for covers . . . only to kick them away again when the fever returned. Slowly, the patient's skin turned yellow and patches of the inside of his mouth began to ooze blood. Nausea came and passed, and returned. There was a pan . . . by the bed to catch the black vomit, a mixture of blood and digestive juices. Two thirds of the patients eventually recovered . . . and became immune. For those who didn't live the jaundiced skin became yellower and yellower. The end was near when . . . tests detected that protein had begun to leak out of the blood, through the kidney membranes and into the urine. Shortly after that, the kidneys shut down, and the flow of urine ceased. When the kidneys died, so did hope. The wracking hiccups began. . . . If the patient was lucky, he went into a coma at about that time. If he was not so fortunate, consciousness faded into delirium and he screamed and cried out in his living nightmare until just before death, which usually occurred between the sixth and the ninth day."

Origins

It appears that yellow fever did not exist in the Western Hemisphere before Christopher Columbus arrived in the New World. Indeed, the first recorded yellow fever epidemic in the Western Hemisphere was in 1648 in Yucatan, Cuba, and Barbados. By the 1690s, it was present in North America, especially Charleston, New York, and Philadelphia—port cities with dense populations that provided the means for effective transmission. There was an outbreak in Boston in 1693 after the arrival of the British fleet from Barbados. Epidemics broke out in Philadelphia and Charleston in 1699 and in New York in 1702. Yellow fever was absent in the British colonies between 1763 and 1792; however, in 1793 there were outbreaks in the West Indies and Haiti and on the island of Santo Domingo. In a 3-month period, 44% of the British soldiers from the 41st Foot Regiment and the 23rd Guard died in Santo Domingo. Refugees from these disease-ridden areas carried yellow fever to several ports, and one of these, Philadelphia, was particularly hard hit during that summer: 5,000 out of a total populations of 60,000 died. At the time Philadelphia was the U.S. capital, and the political leaders George Washington, John Adams, Thomas Jefferson, and Alexander Hamilton were witnesses to the devastating plague. By September the federal government was forced to shut down when six clerks in the Treasury Department contracted yellow fever, five others fled to New York, and three workers in the post office and seven officers in the Customs Service became ill. George Washington escaped to Mount Vernon and recommended that the clerks and the entire War Office move out of Philadelphia.

Although a quarantine was put into effect, it failed to stop the spread of yellow fever, and the public health authorities concluded that the disease was not imported but was the result of the miasma that rose up from rotting coffee on the wharf and garbage in the streets. Alexander Hamilton, who at the time was the Secretary of the Treasury, came down with yellow fever and left Philadelphia, but after being refused entry into New York City he and his wife traveled to Green Bush in upstate New York to stay with his wife's father. In Green Bush, he and his wife were forced to remain under armed guard until their clothing and baggage had been burned and their carriage had been disinfected.

Yellow fever spread across the globe, moving from port to port, on oceangoing vessels: it was in Brazil in 1686, Martinique in 1690, Cadiz, Spain in 1730, and later Marseilles, France, and Swansea, Wales (1878). Knowing that the disease was contagious (but unaware of its actual cause), the staff of Greenwich Hospital in England segregated yellow fever patients, and to forewarn others in the hospital they were dressed in jackets with a yellow patch. The patients were nicknamed "Yellow Jackets," and when the disease broke out on board ship during an extended voyage, the British hoisted a yellow flag—"Yellow Jack"—signifying that the ship was quarantined because an infectious disease was present. Why the yellow color? It is thought that because in the international flag code yellow stands for the letter Q, the color yellow was used to designate a quarantine.

Yellow fever came to be known as Bronze John or the Saffron Scourge because of its telltale symptom: jaundice. During the 1800s there were 39 epidemics in New Orleans—a city called "Bronze John's port of entry." After the 1820s the disease was confined to areas south of the Mason-Dixon line. The North was no longer troubled with yellow fever from about the time that slavery was eliminated. However, in New Orleans, the major port of entry from Latin America, almost every summer between 1790 and the Civil War there were epidemics. Since the cause and the means of transmission of yellow fever were unknown, it was accepted as a way of life in the American South. The peak of yellow fever epidemics in the United States occurred in 1850, when there was an intense debate over slavery. Some feared an uprising of the immune blacks; others suggested that yellow fever was God's punishment of the whites who engaged in slavery, whereas others claimed that slavery was good for the southern plantation system since the deaths due to yellow fever were less catastrophic for black laborers. Fear of the Saffron Scourge drove many people from the coastal communities, and those who could afford to do so left the South during what was called the summer "sickly season." Yellow fever did not occur in winter in parts of the American South where winters were cold and frosty. However, each summer it was reintroduced from Latin America and Africa. In the South, yellow fever had an interesting pattern: (i) it was confined to cities; (ii) it affected newly arrived immigrants and northerners, because they lacked the immunity of the local residents; and (iii) Blacks, demonstrating a natural immunity, were rarely affected. As a consequence, yellow fever had its major impact on the urban, not the rural, South.

The epidemic nature of yellow fever, with its high mortality and dramatic symptoms, attracted more attention than did other diseases, and it had great impact in the northern press. Consequently, the negative impressions of the South by many in the North were reinforced, leading to a lack of migration from North to South, and there was an increased level of absenteeism in the South during the epidemic season of the summer. Yellow fever was bizarre, exotic, and dreadful, and it was restricted to the South. Yellow fever (along with malaria, hookworm, and pellagra) contributed to the image of the American South as a region that, along with its people, was distinctly different from the rest of the country. Only after these diseases were eradicated—by the mid-1900s—would that perception change.

Failure To Control and the Louisiana Purchase

In 1803, the United States purchased Louisiana from the French for about $15 million. The area was 827,987 square miles and included all the land between the Mississippi River and the Rocky Mountains and from Canada to the Gulf of Mexico. Today it includes 15 states. In 1682 the French explorer Robert La Salle led 50 men from the Great Lakes region down the Mississippi River and claimed the entire Mississippi Valley for France. He named the region Louisiana after his king, Louis XIV. Louisiana became a French colony, and by 1718 New Orleans was its capital. The French were disappointed in the revenues from the colony, so in 1762 they ceded the trading rights to Spain. After several skirmishes between the French and Spanish settlers, Spain took firm control in 1769. Sugar planting and processing began in 1795, and by 1800 it was the major cash crop. Louisiana prospered from sugar cane.

During the American Revolutionary War, Spain allowed the Continental Congress to use New Orleans as a supply base for moving goods up the Mississippi River to the struggling colonies. In 1800, France secretly persuaded Spain to return Louisiana. This troubled the new president, Thomas Jefferson, since Spain had allowed the Americans to deposit their goods duty free in New Orleans and then export them to Europe and elsewhere. At about this time there was a rumor that Spain also planned to give up parts of its colonies, including Florida, to France. If this were carried out, the fear was that the French would cut off American access to the Gulf of Mexico, especially New Orleans. In 1801, Jefferson's Secretary of State, James Madison, learned of the deal between Spain and France, but exactly how much land was to be transferred was not known. The United States tried to block the transfer, but Napoleon Bonaparte turned the United States down. Then in 1802 the Spanish governor, probably under pressure from Napoleon, suspended the right of deposit for American goods in New Orleans. This put the United States on the brink of war with France, and the United States threatened to send 50,000 troops to New Orleans to take the port by force. Instead a land deal was effected, and the reason for averting war was yellow fever.

In 1697 the French had established a colony on the western side of Haiti, with the Spanish taking the eastern side. Haiti was a major sugar producer largely because this labor-intensive crop used slave labor imported from Africa. In 1791 and 1794 there were slave revolts, and one of the leaders, Francois l'Ouverture—himself the son of slave parents—declared himself Governor of Haiti. It was clear to Napoleon that he had to assert his authority, but Haiti was critical for another reason—it would be a staging area for Napoleon's invasion of Louisiana and ultimately the establishment of Napoleon's North American Empire. This of course would interfere with the British interests in North America. In late 1801 Napoleon dispatched 20,000 well-equipped soldiers to Haiti to subdue the rebellion under the command of his brother-in-law General Victor Emanuel Le Clerc. The French landed in early 1802 and, after a few battles, were in control. But there were still minor skirmishes with rebel forces, and Le Clerc was concerned by the number of sick—there were already 1,200 in the hospital, and by April the numbers had increased. The cause of the illness was yellow fever, and soon one-third of the force was ill. By June the French were dying at a rate of 30 to 50 per day. In October Le Clerc himself died of yellow fever. By 1803 the disease had killed 20,000 additional replacements and the French gave up. Of all the French dispatched to Haiti, only 3,000 were left alive, and it is estimated that 50,000 of their comrades died. On 11 April 1803, Napoleon told his finance minister Tallyrand that he could ill afford an American campaign and he feared the British might seize the territory or that they would form an Anglo-American alliance against him. Napoleon renounced all claim to Louisiana. The Louisiana Purchase treaty was signed on 2 May 1803.

The colony of Haiti had been one of France's richest, and in 1798 it had 520,000 inhabitants, but by 1804 not a single white person was left and the total population consisted of 10,000 mulattos and 230,000 blacks. Why were the French decimated by yellow fever? (i) They were a naïve population. (ii) The indigenous population was immune and served as a reservoir for the disease. (iii) The climate was favorable for transmission; the period from 1802 to 1803 was particularly rainy. (iv) Most of the cities had been burned to the ground during the rebellion, and so hospitals and medical supplies were unavailable. (v) The hot, humid climate stressed the French. (vi) The French remained in the low-lying areas, where transmission was favored. Perhaps most importantly, at the time no one knew what caused yellow fever or how to stop its spread. The solution to the problem of transmission would require another century and result from the experiments conducted by Walter Reed and the Yellow Fever Commission.

"Catching" Yellow Fever

Until 1900, no one knew how a person caught yellow fever. Although nurses and others who had intimate contact with yellow fever patients rarely came down with the disease, it was considered to be communicable and to be spread by contact with clothing, bedding, or other inanimate objects. As a consequence, most people—including physicians—were fearful of catching the disease through contact with the clothing and bed linen of patients.

Beginning in 1802, Stubbins Ffirth, a medical student at the University of Pennsylvania, convinced that yellow fever was not contagious, attempted the boldest (and some would say the most foolish and dangerous) of experiments. He carried out self-experimentation to provide for "a revision and alteration of the quarantine laws, which now unnecessarily impede commerce, destroy exertion, injure agriculture, and put manufacturers to many inconveniences." Although he slept all night in a bed covered with black vomit from a yellow fever patient, introduced black vomit into a self-inflicted wound, dropped black vomit into his eye, inhaled its vapors, and swallowed it, he never contracted yellow fever. In 1804 he wrote, "it is doubtful . . . it is communicated from one person to another, and certainly never by means of contagion." Despite this, Ffirth's work had no impact on medical practice in Philadelphia since it only revealed how yellow fever was not spread, not how it was spread, and it did not provide a way for people to protect themselves from the Saffron Scourge. Indeed, Benjamin Rush, a leading physician in Philadelphia, advised that when yellow fever came the best course of action was to leave the city and travel to the countryside where the air was clear. He said, "There is only one way to prevent the disease—fly from it."

Because yellow fever was endemic in Cuba, it posed a danger to all countries with which Cuba traded. Furthermore, if the United States was intent on taking over the failed Panama Canal project from the de Lesseps Company, it would need a "yellow-fever free" Cuba as a base of operations. In 1898 when the Spanish-American War broke out and it appeared that fighting would take place on Cuban soil, yellow fever became a primary concern of the military. In 1900, the U.S. Army sent a four-member team—the Yellow Fever Commission—to Cuba to investigate its cause. The head of the Commission was Major Walter Reed. Their first task was to confirm or reject the claim made (in 1897) by the Italian pathologist Giuseppe Sanarelli: yellow fever was caused by *Bacillus icteroides*, literally "yellow jaundice germ."

Sanarelli was an intrepid and capable microbe hunter who had studied at the Pasteur Institute in Paris. Working in Brazil and Uruguay, he found a bacterium in the blood of individuals suffering from yellow fever. Following Koch's postulates, he cultured the "germ" in the laboratory and then injected it into five South American subjects; three died with jaundice. His work was hailed as a milestone, and awards were heaped upon him for his discovery. However, others were less enthusiastic about his methods since he had infected the human subjects without their permission. The foremost physician of the time, William Osler, was so distressed that he wrote, "To deliberately inject a poison of . . . high virulency into a human being, unless you obtain that man's sanction, is not ridiculous, it is criminal."

Examination of blood from yellow fever victims by the Commission (and again following Koch's postulates) determined that *B. icteroides* was not the cause of yellow fever and was nothing more than the "hog cholera" bacillus; its appearance in Sanarelli's cultures was due to sloppy technique—he had forgotten to wash his hands and had contaminated the blood samples. The Commission then turned its attention to the hypothesis of the Cuban-born physician Carlos Finlay; Finlay trained at Thomas Jefferson Medical College in Philadelphia and had experienced the disease in that city during the 1853 outbreak as well as in Cuba. He graduated in 1855, set up a practice in Havana in 1857, and in 1881 proposed that a mosquito transmitted yellow fever. Over a period of 19 years, Finlay conducted 104 experiments in which he was unsuccessful in infecting humans through the bite of female *Aedes* mosquitoes. Thus, his proposal about the mode of transmission required a great leap of faith. He intuited that since yellow fever affects the blood capillaries, a blood-sucking intermediate must be responsible for transmission! (In fact, there was nothing wrong with Finlay's theory; it was simply a matter of timing: the mosquitoes he used did not transmit the infection because, after mosquitoes had fed on a yellow fever subject, he did not allow enough time to elapse before these mosquitoes were allowed to bite a volunteer.)

Despite a lack of experimental evidence, Finlay's belief was consistent with the work of others who had demonstrated that "blood suckers" could act as disease vectors. In 1878, Patrick Manson discovered mosquitoes to be transmitters of elephantiasis; in 1892, Theobald Smith and Frederick Kilbourne showed that ticks spread Red Water Fever in cattle; in 1894, David Bruce demonstrated that African sleeping sickness was transmitted by tsetse flies; and in 1897, Ronald Ross, acting on Manson's suggestion, discovered the mosquito to be the vector of malaria. This was the background when the Yellow Fever Commission began its work. At the time no animals except humans were known to be susceptible to yellow fever; therefore, human volunteers were needed to test Finlay's hypothesis. The volunteers were paid $100 for participating, and another $100 bonus if they contracted yellow fever. Mosquitoes (*Aedes aegypti*) were reared from eggs provided by Finlay to ensure that the mosquitoes had not been previously exposed to a yellow fever patient or humans with some other blood disease (such as malaria). The first experiment involved nine human subjects who were bitten by mosquitoes almost immediately after having fed on a patient with yellow fever; none became ill. Next, one of the members of the Commission (James Carroll) volunteered to be bitten by a mosquito that had fed on a yellow fever patient 12 days earlier. Two days later he experienced the early signs of yellow fever, and 4 days later had full blown disease with fever, headache, swollen gums, and a yellowing of the skin. Since no malaria parasites were found in his blood, it was concluded that he had become infected by having the germ of yellow fever introduced into his body by a mosquito bite. Carroll recovered; however, since he had been in contact with yellow fever patients a few days before his illness, it was possible that the mosquito bite had not caused his disease but was an incidental factor. Therefore, another member of the Commission (Jesse Lazear) and a young soldier (Private William Dean) who had not been in contact with patients allowed themselves to be bitten. Five days later Lazear began to feel ill, the disease progressed rapidly, and on day 12 he died. The soldier also came down with yellow fever but did not die. Just 2 months later, and working from a sample of three, Reed reported, "The mosquito serves as the intermediate host for the parasite of yellow fever, and it is highly probable that the disease is only propagated through the bite of this insect." Reed's conclusions were greeted with derision, and editorials in newspapers ridiculed the mosquito hypothesis. The 2 November 1900 *Washington Post* wrote, "Of all the silly and nonsensical rigmarole of yellow fever that has yet found its way into print—and there has been enough of it to build a fleet—the silliest beyond compare is to be found in the arguments and theories generated by the mosquito hypothesis."

To satisfy his critics, Reed returned to Cuba in late November, determined to provide definitive proof that the mosquito, and not filth, was responsible for the transmission of yellow fever. To test the filth hypothesis, seven volunteers were placed in a small house with screened windows and doors to exclude the entry of mosquitoes. The volunteers were then exposed to bedding and clothing soiled by the bloody vomit and stools from yellow fever victims. They also slept in pajamas and other clothing that had been worn by those who had died from yellow fever. After 63 days, none had become sick. To test for mosquito transmission, another house was used. This was partitioned into two areas by a fine-mesh wire screen that would not allow mosquitoes to move from one area to the other. In one area 15 mosquitoes that had previously bitten yellow fever patients were released and allowed to bite volunteers; volunteers on the other side of the screening were not exposed to mosquitoes. The volunteers were then removed to screened tents and were monitored for disease. Only the volunteers exposed to the mosquitoes developed yellow fever. On 15 December 1900, Reed contacted the Surgeon General: "Theory conclusively proved." There was more than proof of a theory by the Yellow Fever Commission; it led to an effective means of controlling the disease: William Gorgas, Chief Sanitary Engineer in Havana, was able to introduce antimosquito measures that decreased the number of yellow fever cases from 1,400 in 1900 to none in 1902.

The Yellow Fever Commission identified the vector for yellow fever and made possible public health measures to control its spread; it also formalized the covenant between medical investigators and human volunteers. The conduct of their human experimentation was guided by ethical responsibilities, was the forerunner of the current practice of informed consent, and was used in the formulation of the Nuremberg code. (It is regrettable that informed consent has not always been applied when humans have been used as guinea pigs, e.g., the Tuskegee Syphilis Study.)

The experiments conducted by the Yellow Fever Commission clearly showed that for natural transmission to occur, the female *Aedes* mosquito must feed on a yellow fever victim within the first 3 or 4 days of illness and then the virus must be incubated for 10 to 12 days in the mosquito before the insect is infective. The mosquito is then infective for the remainder of its life. After being bitten by an infected mosquito, it takes 3 to 6 days before fever occurs.

The *Aedes* mosquito, native to Africa, is a highly domesticated species that breeds in small bodies of still water—water casks, cisterns, or tin cans are excellent sites for it to lay its eggs. It crossed the ocean aboard ships, entered port cities, and established itself in areas around the world where the temperature always stayed above 72°F. Yellow fever, endemic in Africa, traveled aboard trading ships with their cargoes of slaves. These black

Africans, though infected, were resistant to the disease, and their blood served as a source of infection for the mosquito. This peculiarity of mosquito-human association meant that mosquitoes carrying yellow fever could remain aboard for weeks or months; on land where there were cases of yellow fever, the mosquito would act as an efficient vector. Yellow fever can become endemic only when the climate is warm enough to permit year-round mosquito activity.

Successful Control and Construction of the Panama Canal

The British Empire existed between the reigns of two great queens, Elizabeth I (reigned 1558 to 1603) and Victoria (reigned 1837 to 1901). The domination was based in part on Britain's ability to rule the waves by controlling strategic points around the globe. These ports served not only for provisioning, coaling, and repair of ships but for military ventures as well. The first of these was Gibraltar; later, Capetown, Hong Kong, Singapore, Ceylon, the Falklands, Aden, and Suez were added. Britain held the Mediterranean, Red Sea, and Indian Ocean in its control, but it let go of the Isthmus of Panama, not because of fortresses or cannons but because of a mosquito and a virus.

The success of Ferdinand de Lesseps in building the Suez Canal (1859 to 1869) encouraged the French to attempt to build a Panama Canal. (The Suez Canal—connecting the Mediterranean Sea and the Red Sea—shortened the route between England and India by 6,000 miles, yet the canal itself is only 100 miles long.) The French raised funds for de Lesseps' "Compagnie Universelle du Canal Interoceanique," acquired the land rights to build the Panama Canal from Colombia, and also gained title to the Panama Railroad (which had been built in 1855 by a group of New York executives at a cost of $5 million) for $20 million. They began digging in 1882, but 7 years later, after digging out 76 million cubic yards of earth, they gave up due to mismanagement, poor skills, theft, and disease. The French lacked the knowledge about the mode of transmission of yellow fever, so they inadvertently contributed to its spread by their use of large pots with stagnant water in gardens and under the legs of the barracks and hospital bed to retard trafficking by ants. These water vessels were exceptionally good places for *Aedes* to breed; by 1882 there were 400 deaths among the French and European engineers and laborers, and in the next year there were 1,300 deaths from yellow fever and malaria. Indeed, it is estimated that at any given time one-third of the workforce was sick with yellow fever; in 1884, with more than 19,000 laborers, 7,000 were ill. During the time the French managed the Panama Canal construction project, more than 22,000 workers died. By February 1889, de Lesseps' Compagnie Universelle du Canal Interoceanique went bankrupt.

The California gold rush stirred U.S. interest in a canal connecting the Atlantic and the Pacific as early as 1849, and an interoceanic canal became even more enticing during the Spanish-American War of 1898 because of the difficulty of sending ships from San Francisco to Cuba to reinforce the Atlantic fleet. (The distance from New York to San Francisco with the canal is 5,200 miles; the distance without the canal is 13,000 miles because ships have to go around South America.) In 1899 Congress authorized a commission to negotiate a land deal, and the French sold their rights and the railroad for $40 million. However, Colombia refused to sign a treaty involving a payment of $10 million outright and $250,000 annually because the figure was considered to be too low. A group of Panamanians feared that Panama would lose the commercial benefits of a canal, and the French worried about losing its sale to the United States, so with the backing of the French, the Panamanians, encouraged by the United States, began a revolution against Colombia. The United States, in accord with its treaty of 1864 with Colombia to protect the Panama R.R., sent in troops. The Marines landed at Colón and prevented the Colombian troops from marching to Panama City, the center of the revolution. On 6 November 1903, the United States recognized the Republic of Panama, and 2 weeks later the Hay-Bunau-Varilla Treaty was signed, giving the United States permanent and exclusive use of the Canal Zone (10 miles wide, 50 miles long). Dealing directly with Panama, the United States paid $10 million outright plus $250,000 annually beginning in 1913. The United States also guaranteed Panama's independence.

Construction began in 1907, and the canal was completed in 1914. The project removed 211 million cubic yards of earth. More than 43,400 persons worked on the canal, most of whom were blacks from the British West Indies. The cost of construction was $380 million. In 1936, the U.S. payments to Panama increased to $430,000; by 1955, the U.S. payments to Panama had increased to $1,930,000; and in the 1970s they had increased to $2,328,000. In the 1960s there were riots, and other concessions were granted. Panama continued to exercise greater and greater control over the canal and took complete control on 31 December 1999.

The building of the Panama Canal was seriously affected by the problems of mosquito-borne disease, principally yellow fever. In March 1901,

Major William C. Gorgas, the Sanitary Engineer on the Canal Project, followed up his success in Havana by initiating measures to eliminate mosquitoes and their breeding sites. During construction of the canal, all patients with yellow fever were kept in screened rooms. Workers were also housed in copper-screened houses. Drainage and kerosene spraying were the mainstays of mosquito control. The magnitude of Gorgas' accomplishment can be best appreciated by the number of deaths. During the canal days of de Lesseps, the death rate was 176/1,000 people. When the canal was completed by the United States, the death rate from all causes was 6/1,000. The last fatal case of yellow fever occurred in Panama in 1906.

A Vaccine

Although the Yellow Fever Commission found no evidence to support Sanarelli's claim that *B. icteroides* was the cause of yellow fever, it was still possible that another microbe could be responsible. However, that microbe could not be a bacterium since none could be seen when the blood of a yellow fever victim was examined under the microscope. The first clue to the nature of the germ came when a member of the Commission (Carroll, a microbiologist) took a sample of blood from a patient during the first 3 days of illness and injected this blood into healthy volunteers and they came down with yellow fever. The same result occurred when the blood was filtered through a Chamberland filter. Thus, it was clear the germ of yellow fever had to be smaller than a bacterium.

Twenty-five years after the work of the Yellow Fever Commission, the germ of yellow fever was shown to be a virus. In 1927, Adrian Stokes, Johannes Bauer, and Paul Hudson of the Rockefeller Foundation found that blood from a person (named Asibi) from Ghana, who was suffering with a mild case of yellow fever, was infective for the common Indian rhesus monkey. This eliminated the need for human volunteers and provided an animal model of the disease. Shortly thereafter, Max Theiler, a South African working in the Rockefeller Foundation laboratories in New York, found that the common white mouse was susceptible to the disease if the animals were inoculated with virus into the brain. The disease in mice was different from that in humans since it caused an encephalitis and did not involve the kidneys, liver, or heart. By serial passage in mice, Theiler observed that the virus became less pathogenic for monkeys and after many passages it became stable or, to use the term first coined by Pasteur in his work on rabies, fixed. Furthermore, mice were protected from a lethal dose of virus if they first received serum from humans or monkeys that were immune to yellow fever. Although this method might have provided the basis for a vaccine, it was cumbersome and impractical and could not be used for large-scale immunization; therefore, Theiler turned his attention to growing the virus in tissue culture containing minced chicken embryo tissue with small amounts of nervous tissue. In 1937, after many passages in culture, Theiler found the virus had mutated and lost its ability to produce a fatal encephalitis in rhesus monkeys as well as mice. This attenuated virus, called the 17D strain, when injected into monkeys resulted in their producing antibodies which were protective against the normally lethal Asibi strain. Next, human volunteers—personnel working in Theiler's laboratory—were vaccinated with the 17D strain, and although there were mild side effects they made antibodies that were able to neutralize the virus. Clearly, the 17D strain could become a practical protective vaccine for yellow fever. Indeed, with stepped-up production of 17D, millions have been successfully vaccinated against yellow fever. For this work, Max Theiler received the 1951 Nobel Prize for "discoveries concerning yellow fever and how to combat it."

Since vaccination against yellow fever using the attenuated 17D strain is so effective and safe, it has been suggested that some of its successful traits could be applied to vaccines against other diseases. Work is under way to piggyback the immune stimulating features of other vaccines onto the yellow fever shot. The hope, and it is only a hope at present, is that by riding on yellow fever's coattails—and using chimeric 17D vaccines—it will be possible to generate potent immunity against malaria, HIV, and possibly some cancers.

Consequences

Yellow fever virus evolved from other mosquito-borne viruses about 3,000 years ago, probably in Africa, where it remains endemic. It was imported to the New World and elsewhere aboard trading ships with their cargoes of human flesh, i.e., slaves. Although the African slaves were easily infected, their longstanding association with the disease allowed them to resist its effects, so that fewer died from the infection than did Caucasians, Amerindians, or Asians. It is a bitter irony that as smallpox and measles devastated natives along the Caribbean coast and islands, growing numbers of African slaves were brought in to replace those plantation laborers.

When the value of black Africans over natives became apparent by virtue of their resistance to yellow fever, the importation of black Africans increased further, leading to the major scourges in the 18th and 19th centuries in colonial settlements in the Americas. Systematic investigations of transmission and identification of the yellow fever virus in the 20th century have resulted in prevention of the disease by vector control as well as the development of a protective vaccine; however, the disease remains a continued threat to those who live and travel to regions where yellow fever is endemic if they have not been vaccinated.

10

The Great Influenza

The young American soldier had been dispatched to fight in the "war to end all wars." He never fired a shot. Instead, he lay helpless on a hospital cot, wounded by a force more deadly than that launched by the enemy. It had begun innocently enough: headache, chills, and fever. Pains in his joints prevented him from standing. He became nauseated; he began to sweat, and he vomited. The mucous membranes of his throat and nose were congested, and he had a persistent cough. Lips, ears, nose, cheeks, tongue, and fingers—his entire body began to turn an indigo blue color. The stench of corruption filled his nostrils; it was the sweetish sickening smell of rotting flesh. It was the breath of death. Suddenly blood began to pour from his body—at first a trickle from the nose, mouth, and ears and around the eyes, and then it gushed. His lungs were being ripped apart. At autopsy the doctor said the soldier's lungs resembled "melted currant jelly."

The disease that killed the young recruit that day in the fall of 1918 would, over the next 2 years, kill another 22 million. They were all killed by influenza, for which there was no cure. Influenza is an ancient disease that has caused worldwide outbreaks at irregular intervals through recorded history. Over the past 300 years at least 10 and perhaps 20 global epidemics—pandemics—have been interspersed with a larger number of milder and more localized outbreaks. In 1889 the Russian flu killed 1 million, in 1957 the Asian flu killed 2 million, and in 1968 1 million died from the Hong Kong flu. The 1918 flu, however, was different: the healthiest and strongest seemed the most vulnerable, it killed more people than the Black Death of the Middle Ages and more than AIDS over the past 25 years, and it did so quickly. It moved with great speed across the continents, leaving death and disorder in its wake. Ironically, the media and public officials fueled the panic and terror so prevalent in 1918 to 1920, not by their exaggeration of the threat but by minimizing the outbreak and attempting to reassure the people that it was nothing more than "la grippe." Another deadly influenza epidemic is a certainty. The challenge is how to prepare for the unpredictable.

Cause

Hippocrates, the father of medicine, recorded a flu outbreak as early as 412 BC. In 15th-century Italy, when a cold wind was believed to influence an outbreak, the disease was named "influenza di fredo." The first well-recorded epidemic occurred in 1580; it probably began in Asia and then moved on to Africa and Europe, where especially large numbers of deaths occurred in the cities. Over the centuries, influenza epidemics were ascribed to the alignment of the stars or poisonous vapors (miasmas) as well as the weather, but by the 19th century the leading candidate was a bacterium (named *Haemophilus influenzae* by the German microbe hunter Richard Pfeiffer) found in the throats of patients suffering from the disease. However, it would be several decades before the discovery that Pfeiffer's bacterium was not the cause of influenza but simply a bystander.

In 1918 a veterinarian in Iowa made a seminal observation: at the National Breeders Show in Cedar Rapids, Iowa, some swine were ill with a disease that resembled human influenza, which was already raging in many countries. Further, there were other reports from Iowa that epidemics in pigs were sometimes followed by flu in families that raised swine. Outbreaks of swine flu continued in the Midwest in 1922 and 1923, and during this time veterinarians were able to transmit the disease from pig to pig by using nasal mucus. However, they were unable to find the cause of swine flu. In the 1930s, Richard Shope identified the true cause of influenza as a virus. After receiving his medical degree from the University of Iowa, Shope, the son of an Iowa physician and farmer, followed up on the work of the Iowa veterinarians at the Rockefeller Institute in New Jersey. Working with an influenza virus strain that infected pigs, he was able to show that the virus could be transmitted between pigs when he performed his experiments with filtered nasal mucus (thereby excluding Pfeiffer's bacterium since any bacteria present would be too large to pass through the pores of the filter and would not be in the filtrate although the much smaller viruses could move through the pores into the filtrate).

But was this swine virus the same as the one causing human infections? An influenza outbreak in England in 1933 provided Wilson Smith, Christopher Andrewes, and Patrick Laidlaw (working at the National Institute for Medical Research in Mill Hill, London) with an opportunity to isolate the first human influenza virus. Among the many people who came down with influenza were several members of the Institute, including Andrewes. A filtrate from Andrewes' throat washings was placed into the nasal passages of ferrets, the same route Shope had used in his pig experiments, and the ferrets came down with influenza. This was the first evidence that a virus caused human influenza, and it fulfilled Koch's postulates. The strain was named influenza A virus.

The use of ferrets for influenza research was serendipitous. In the late 1800s and 1900s, English country gentlemen and women who were engaged in fox hunting with hounds became concerned about an outbreak of canine distemper—a respiratory disease related to human measles—that resulted in the death of many of the hounds. These sportsmen and -women banded together, and, with the help of subscribers to *Field Magazine,* a journal that catered to fox hunters, they were able to raise enough money to fund research on canine distemper on a farm in Mill Hill where the sick dogs could be isolated and studied. As a result of research at Mill Hill, a protective vaccine for canine distemper was produced in 1928; however, the continued use of dogs for research created problems with pet owners and anti-vivisectionists, and the fact that some dogs were immune because of a previous encounter with the virus also complicated research. When it was discovered that ferrets could be substituted for dogs, these problems vanished. Thus, the ferrets in the Mill Hill laboratory of Smith, Andrewes, and Laidlaw that had been used to study canine distemper virus were available for studies of the influenza virus.

In 1940 a distinctly different strain of influenza virus was isolated from a human; it was named influenza B virus. Subsequent studies showed that both the influenza A and B viruses could be grown in chicken embryos and that infected fluid from such embryos would clump chicken red blood cells. These developments allowed the production of large quantities of influenza virus in the laboratory and provided the starting material for an inactivated vaccine. In addition, because the clumping reaction could be inhibited by specific antibodies in the serum of humans or animals infected or vaccinated with influenza viruses, different strains could be distinguished (typed), much as we are able to distinguish the different types of blood (Rh, M, N, A, B, AB, and O) in humans.

The influenza virus is so tiny that 10,000 would occupy less space than the period at the end of this sentence. Since it is small and light, the virus is easily spread in an aerosol mist by coughs and sneezes; each droplet in the contagious mist may contain up to 500,000 virus particles. When the virus is magnified more than 100,000 times under an electron microscope, its outer surface can be seen to be covered with two kinds of projections, resembling spikes and mushrooms. The spikes, made of a protein called hemagglutinin (H), are interspersed with mushroom-like protrusions composed of another protein, neuraminidase (N). The job of H is to act like grappling hooks to anchor the influenza virus to host cell receptors made of sialic acid. (The reason why infected chicken embryo fluid clumps chicken red blood cells is that the H on the virus surface hooks together several sialic acid-containing red cells to form a network.) After binding, the virus enters the host cell, replicates its genetic material (RNA), and produces new viruses. The emerging viruses are coated with sialic acid, the substance that enabled them to attach to the host cell in the first place. If the sialic acid were allowed to remain on the virus and on the host cell, these new virus particles with H on their surface would be clumped together and trapped much like flies sticking on flypaper. The N allows the newly formed viruses to dissolve the sialic acid "glue"; this separates the viruses from the host cell and allows them to plow through the mucus between the cells in the airways along the respiratory tract and to move from cell to cell. The entire process—from anchoring to release—takes about 10 h, and in that time 10^5 to 10^6 viruses can be produced.

Drifting and Shifting Thwart Control

Control of cholera depends on the separation of sewage from drinking water; control of malaria depends on the eradication of mosquitoes and treatment of the infected individual; control of viral diseases such as measles, mumps, smallpox, polio, and yellow fever depends on the isolation of the virus and production of a protective vaccine. But influenza is different. Seventy years after the isolation of the causative agent and the development of vaccines, influenza remains the only infectious disease that appears periodically in life-threatening pandemics. If a virus never changed its surface antigens, as is the case with measles and mumps, the body could react with an immune response—antibodies and cell-mediated immunity—to the foreign antigens during an infection or by a vaccine, and there would be long-lasting protection. Indeed, if a person encounters the same virus a second time, the immune system, having been primed, swiftly eliminates that virus and infection is prevented. The persistence and unpredictability of the influenza virus result from its ability to change its surface covering—acquiring a new cloak, as it were—and thereby becoming invisible to the body's immune system. As a result, this notorious viral shape-shifter is able to swiftly travel across the globe, leaving millions of sick and dying in its wake.

Influenza viruses contain eight separate RNA segments (genes) that code for 10 proteins. The two types of influenza virus—group A and group B—differ from one another with respect to the six genes that do not encode the H and N proteins. Group A influenza viruses infect pigs, horses, seals, whales, and birds as well as humans, whereas group B viruses infect only humans.

A pandemic is the viral equivalent of the perfect storm. For such to occur, three unlikely and unpredictable conditions have to be met: First, a "new" influenza virus must emerge, i.e., one for which a person has no immunity. Second, the virus must be pathogenic, i.e., must cause disease. Finally, the virus must be easily transmissible—through coughing, sneezing, or even a handshake. The emergence of a new virus stems from the capacity of the virus to change the shape of its H and N proteins so that the immune system is unable to recognize the new surface contours. As a result, antibodies are unable to bind, and therefore this virus cannot be neutralized. This is the reason why a vaccine that was produced against an influenza virus with a characteristic surface (fingerprint) and that was protective last year may be ineffective next year: if the new fingerprint is unrecognizable, the vaccine does not work.

The sites where surface alterations occur in H and N (and recognized by a specific antibody) are called antigenic sites. A small change in the antigenic site is called antigenic drift. Drift occurs because the RNA copying mechanism of the virus is faulty and the mistakes made are not corrected because the virus' RNA polymerase is not a very good proofreader. These errors (i.e., mutations) in RNA segments 4 and 6 allow the virus to change its surface H and N proteins swiftly. Influenza B viruses can only undergo antigenic drift, and as such they are responsible for regional interpandemic flu outbreaks.

Influenza A viruses can change their fingerprint in a much more dramatic way, and this is called antigenic shift. Type A viruses undergo antigenic shift every 20 to 30 years and are responsible for pandemics. One investigator has said that if antigenic drift is likened to a shudder, antigenic shift is more like an earthquake. An antigenically shifted "new" virus is so different from other human influenza viruses that it could not have arisen by mutation. The new virus has its origins in wild aquatic birds such as ducks and other kinds of waterfowl, where they cause no apparent harm. The viruses replicate in the cells lining the intestinal tract of the waterfowl, and high concentrations of virus are eliminated in their feces. Because such birds can migrate thousands of miles, they can spread the virus across the globe even before it enters the human population. Under most circumstances, the influenza virus in wild birds does not replicate well in humans and so must move to an intermediate host—usually domestic fowl or pigs—that ingest the virus by drinking water contaminated by fecal material from the waterfowl. The domestic fowl tend to be dead-end hosts because most sicken and die, but pigs live long enough to serve as "virus mixing vessels" in which the genes of influenza viruses from birds, pigs, and humans can be mixed because pig cells have receptors for both bird and human viruses. It is by this coming together of virus genes (called reassortment) that new strains of influenza virus are produced. If two different viruses—one avian and one human—infect the same cell, a random shuffling of the eight gene segments can take place and yield up to 256 (or 2^8) different offspring. This sharing of genes leads to immediate and dramatic changes in the viral H and N proteins and possibly other gene products as well.

There are 15 different subtypes of H and 9 different subtypes of N, all of which are found in avian influenza viruses. The subtypes of type A viruses are named according to the particular variant of the H and N proteins they contain and are designated by a letter and a number. Only three H subtypes—H1, H2, and H3—and two N subtypes—N1 and N2—have been established in human infections; only one subtype of H and N is found in influenza B viruses. The strain that caused the 1918 pandemic was H1N1, and in 1976 a swine influenza virus (H1N1) was isolated from military recruits at Fort Dix, one of whom died. This latter mini-epidemic not only confirmed the work of Shope (implicating swine viruses in human disease) but also indicated that pigs serve as a major reservoir for these viruses. Recent evidence shows that the influenza found in swine in the 1920s and 1930s, and which reappeared in 1976, did not come from birds or other swine but from infected humans!

The pandemics of influenza that have swept across the world in the last 50 years appeared first in China and then in Hong Kong; this is because birds, pigs, and humans live in close proximity in China. In 1957 a new virus, the Asian flu, having two new surface proteins—H2 and N2—appeared; since H2 had 66% similarity in amino acid composition to H1 and since N2 had 37% similarity to N1, the new virus must have arisen by antigenic shift. Thus, between 1918 and 1957—39 years—a time when there was little or no preexisting immune protection against these novel surface proteins, their appearance resulted in a pandemic. (All six other viral genes remained unchanged.) Eleven years later, another variant resulted in a pandemic, but in this instance only the H mutated (H2 to H3). In this pandemic there was a lower mortality than in 1957 (when 70,000 people in the United States and many more worldwide died), presumably because the lack of change in N provided some protection to persons previously exposed to H2N2. The 1997 Hong Kong influenza outbreak (H5N1) was caused by an avian influenza virus that acquired its virulence by reassortment of genes from geese, teal, and quail housed together in poultry markets. Eighteen people acquired the infection from contact with the feces of infected chickens, not from other people. Fortunately the spread to other humans was halted before person-to-person (mucus droplet) transmission could result in a full-blown flu epidemic because health authorities enforced the slaughter of more than a million fowl in Hong Kong's markets. Had this not occurred, it is likely that one-third of the human population would have sickened and died.

In 2001 a new variety of the H5N1 virus appeared in the poultry markets of Hong Kong. No human cases occurred because thousands of fowl were destroyed before transmission occurred. Then in 2004 another kind of H5N1 appeared. This H5N1 virus was unusual in that it did not require an intermediary host: the virus jumped directly from birds to humans. This strain has killed hundreds of millions of fowl as well as mice, pigs, cats, and tigers in nearly a dozen Asian countries. It has also killed people: 76 people have died, and this represents half of those infected.

In January 2006, the H5N1 virus felled 1,300 birds in eastern Turkey. By February it was in France—the first European country to suffer an outbreak. On 5 February a russet-headed duck was found dead in a pond in eastern France, and 5 days later another migratory duck, of a different species, was found dead from H5N1. The virus appears to have arrived in France with ducks that migrated from the Black Sea to escape the unusually cold weather. Soon after H5N1 arrived on a French poultry farm, 11,000 turkeys were suffering from diarrhea, and a week later 430 were dead. The threat to European fowl (and perhaps humans) will continue as pintails, garganeys, and shoveler ducks begin arriving from Africa on their annual northward migration. Although migratory waterfowl seem to be the principal means of transmission, this is not the only way the lethal virus has moved out of Asia. The pattern of infections from the Far East into Western Europe follows the railway lines more closely than it does the bird migratory pathways. One factor in the spread of H5N1 may be Asian chicken manure, used in everything from fishponds to poultry feed to fertilizer that is spread on the fields, as well as the crates in which the poultry are shipped. A cat on the Baltic Sea island of Rugen in Germany tested positive for the virus; it may have contracted the virus by eating the carcass of an infected bird. Cats can spread the virus to other cats, and although there is no record of cat transmission to humans, it is a theoretical possibility. So far we have been lucky because this bird influenza has not been transmissible from human to human. However, when it mutates into a contagious form among humans, will we be prepared to curb its spread?

Pandemics and Pathogenicity

Although influenza pandemics occurred in 1957 (Asian flu) and 1968 (Hong Kong flu) and scares occurred in 1976 (swine flu) and 1977 (Russian flu), the mother of all pandemics was that of 1918. The 1918 outbreak of influenza remains one of the world's greatest public health disasters. Some have called it the 20th century's weapon of mass destruction. It killed more people than the Nazis and far more than did the two atomic bombs dropped on Japan. In 24 weeks it destroyed the lives of more people than those killed by AIDS in 24 years. As with AIDS it killed those in the prime of their life: young men and women in their 20s and 30s.

The 1918 to 1920 pandemic brought more than human deaths: civilian populations were thrown into panic, public health measures were ineffectual or misleading, there was a government-inspired campaign of disinformation, and people began to lose faith in the medical profession. It also hastened the end of World War I. In the spring of 1918, as a result of the Russians withdrawing from the war, the German army began to mount a massive offensive on the Western front consisting of 1 million men, 37 infantry divisions, and 3,000 guns. In May, when the German army and its artillery were within striking distance of Paris, a German victory over the Allies seemed inevitable. However, in late June, when the German army began to suffer from an outbreak of influenza—2,000 men in each division were afflicted—the tide of war began to turn. The strength of the German army withered because of sickness as well as deficiencies in supplies and food. The Allied army, invigorated by the arrival of American soldiers, mounted a grand offensive that blocked the German advance and regained French ground. Because the pandemic killed more than twice the number who died on the battlefields of World War I, it probably hastened the armistice that ended the "Great War" in Europe. It also played a part in the rise of Adolf Hitler and sowed the seeds for World War II. During the Armistice negotiations, Woodrow Wilson, President of the United States, was so ill and disoriented with influenza (acquired during an outbreak in Paris) that he acceded to the formula proposed by the French President George Clemenceau: Germany would pay reparations and accept full responsibility for the war; there would be demilitarization of the Rhineland; the rich coal fields of the Saar would be mined by the French; Alsace and Lorraine (taken by Germany during the Franco-Prussian War) would be returned to the French. The German air force was eliminated, and its army would be limited to 100,000; Germany would be stripped of its colonies, which would be redistributed among the Allied powers. The harshness toward Germany in the peace treaty helped create the economic hardships, nationalistic reaction, and political chaos that fostered the rise of the Nazi party and ultimately would precipitate another World War.

Where did this global killer come from? Epidemiological evidence suggests that the outbreak was due to a novel form of the influenza virus that arose among the 60,000 soldiers billeted in the army camps of Kansas. The barracks and tents were overflowing with men, and the lack of adequate heating and warm clothing forced the recruits to huddle together around small stoves. Under these conditions, they shared both the breathable air and the virus that it carried. By mid-1918, infected soldiers were carrying the disease by rail to army and navy centers on the East Coast and the South. Then the infection moved inland to the Midwest and onward to the Pacific states. In its transit across America, cases of influenza began to appear among the civilian population.

People in cities such as Philadelphia, New York, Boston, and New Orleans began to ask: What should I do? How long will this plague last? To minimize panic, the health authorities and newspapers claimed that it was "la grippe" and that there was little cause for alarm. This was an outright lie. However, as the numbers of civilian cases kept increasing and when there were hundreds of thousands of infected persons and hundreds of deaths each day, it was clear that this influenza season was nothing to be sneezed at and the peak of the epidemic had not been reached. In Philadelphia undertakers had no place to put the bodies and there was a scarcity of coffins. Gravediggers were either too sick or too frightened to bury influenza victims. Entire families were stricken, with almost no one to care for them. There were no vaccines or drugs. None of the folk remedies were effective in stemming the spread of the virus, and the only effective treatment was good nursing. In cities, as the numbers of sick continued to soar, public gatherings were forbidden and gauze masks had to be worn as a public health measure. The law in San Francisco was that if a person did not wear a mask, that person would be fined or jailed.

Conditions in the United States were exacerbated as the country prepared to enter the war in Europe. President Woodrow Wilson's aggressive campaign to wage total war led indirectly to the nation's becoming "a tinder box for epidemic disease." A massive army was mobilized, and millions of workers crowded into the factory towns and cities, where they breathed the same air and ate and drank using common utensils. With an airborne disease such as influenza, this was a prescription for disaster. The war effort also consumed the supply of practicing physicians and nurses, and so medical and nursing care for the civilian population deteriorated. "All this added kindling to the tinderbox."

The epidemic spread globally, moving outward in ever-enlarging waves. The hundreds of thousands of U.S. soldiers who disembarked in Brest, France, carried the virus to Europe and the British Isles. The infection then moved to Africa via Sierra Leone, where the British had a major coaling center. The dock workers who refueled the ships contracted the infection, and they spread the highly contagious virus to other parts of Africa when they returned to their homes. In Samoa a ship carrying infected persons arrived from New Zealand, and within 3 months over 21% of the Samoan population had died. Similar figures for deaths occurred in Tahiti and Fiji. In a few short years, influenza had spread worldwide and death followed in its wake.

Still to be explained is precisely why the 1918 outbreak of influenza was so pathogenic. It may have had something to do with an exaggerated innate inflammatory immune response with release of lymphokines such as tumor necrosis factor; this can lead to a toxic shock-like syndrome involving fever, chills, vomiting, and headache and ultimately could result in death. Alternatively, perhaps the H1N1 virus that led to the 1918 to 1920 pandemic had a very different kind of H, one more closely related to that of an avian virus, and for this there was little in the way of immune recognition.

Control

Globally, influenza remains an important contagious and incapacitating disease, with 20% of children and 5% of adults developing symptoms. In most cases the disease course is brief (3 to 7 days) and rarely fatal. After an incubation period of 24 to 72 h, fever, chills, headache, aching muscles, and fatigue occur and are followed by a runny nose, sore throat, and persistent cough. After the symptoms appear, the only treatment is supportive. However, on occasion the infection can escalate quickly to produce bronchitis and secondary infections, including pneumonia and heart failure.

The mortality rate—the percentage of those infected who die from their infection—is usually low (<2%). Despite this low mortality rate, influenza can become a major public health problem when there is a large pool of immunologically naïve individuals who may become infected. During the 1918 epidemic, 0.6% of the U.S. population (~675,000) died, and for every person who died in the United States, 50 had the disease. Using today's U.S. population, an outbreak similar to 1918 would kill 2 million.

Each fall we are encouraged to get a "flu shot" so that when the influenza season arrives in winter we will be protected. Scientists at more than 100 World Health Organization laboratories are constantly collecting and analyzing the influenza viruses that are circulating in the human population. After isolating the viruses they identify two type A strains (usually H1 and H3 subtypes) and one type B strain that are most likely to cause an epidemic. Those chosen in the spring are used by vaccine manufacturers to prepare the standard season vaccine for the following winter. However, such a vaccine can protect only against targeted or known strains; it cannot protect against unexpected or unidentified types that may arise after the World Health Organization has made its determinations. Live influenza viruses (unlike the viruses of polio and smallpox used in immunizations) are not used in vaccines, and so immunity, not infection, results. The vaccines that are protective against a known type of influenza virus do not cause disease because, after the viruses have been grown in chicken embryos, they are purified and inactivated by formaldehyde. Whole-virus vaccines are not widely available because they cause adverse reactions in children, whereas those that contain purified surface H or N proteins are well tolerated and activate the immune response. A weakened live virus vaccine may give a stronger and longer-lasting protection, but for such a vaccine it is critical that a return to virulence does not occur. Although these vaccines can be administered as a nasal spray, avoiding the pain of inoculation, they may also produce mild but unpleasant disease symptoms.

If there were a movie called "The Next Pandemic," it would be a creepy thriller. The opening scenes would be somewhere in China, where a virus is being created by shuffling and mixing bits of viral RNA from birds and humans. Because the virus is cloaked in bird surface proteins and has never been seen before by any human immune system, it will remain invisible save for the symptoms it produces in those it infects—fever, headache, nausea, muscle ache, and fatigue. In the beginning, there are a few human cases, and these are restricted to poultry farmers and children in Asia. There is little concern in the rest of the world for a few dead Asians, but in Asia itself control is attempted by destroying large flocks of domestic poultry. The rest of the world continues to take comfort in the belief that there is a species barrier and that this virus is transmissible only to other birds and only rarely infects humans who have come in contact with infected birds. However, when human infections begin to expand beyond Asia and there is no evidence of contact of infected humans with infected birds, there is the realization that this virus has metamorphosed and become more effective at human-to-human transmission. By the time the public health authorities in Asia recognize that poultry culling has not worked and the cluster of human cases has expanded, the virus has hopped aboard intercontinental flights and moved into Australia, Europe, and the United States. As hundreds and then thousands begin to die by drowning as fluid and pus fill their lungs, people begin to panic. They ask: "Who will be next? Can't we quarantine the infected or those most susceptible to influenza? Where are the vaccines and the drugs?" However, there is little in the way of immediate relief. A vaccine cannot be made available for months, and drugs are in short supply. The quarantine is limited and ineffectual. Rioting begins. Hospitals and health clinics are stormed. Pharmacies are robbed. Hospitals become overcrowded with the sick and dying, and there are insufficient numbers of doctors, nurses, and even coffins. Schools, businesses, and transportation systems shut down, but still the numbers of cases and deaths continue to soar. People wear face masks, avoid public gatherings, and are even afraid to greet one another with a handshake. Stock markets fail, and there is economic collapse. Then after 18 months the epidemic wanes. But the toll is enormous: more than 1 billion have sickened and 40 million have died from influenza.

This scenario is fiction, but parts of the film—now retitled "Flu H5N1"—have already been played out in China, Turkey, and France.

Pandemics will continue to plague humankind as long as there is viral mixing, and it may be impossible to prevent infection in those at high risk (i.e., those older than 50 years of age, the very young, the chronically ill, and the immunosuppressed) under any circumstances. In addition, infected persons are at greater risk for pneumococcal pneumonia (a bacterial disease), which annually kills thousands of elderly people in the United States. Influenza combined with pneumonia can result in skyrocketing mortality.

Influenza epidemics cannot be contained by quarantine. The reasons for this failure are easily explained. For an infection to persist in a population, each infected individual on average must transmit the infection to at least one other individual. The number of individuals infected by each infected person at the beginning of an epidemic is given by R_0; this is the basic reproductive ratio of the disease or, more simply, the multiplier of the disease. The multiplier helps to predict how fast a disease will spread through a population. The value for R_0 can be visualized by considering the children's playground game of "touch tag." In this game, one person is chosen to be "it" and the objective of the game is for that player to touch another who in turn also becomes "it." From then on, each person touched helps to tag others. If no other player is tagged, the game is over, but if more than one other player becomes "it," the number of "touch taggers" multiplies. Thus, if the infected individual (it) successfully touches another (transmits), then the number of diseased individuals (touch taggers) multiplies. In this example the value for R_0 is the number of touch taggers that result from being in contact with "it."

The longer a person is infectious and the greater the number of contacts that the infectious individual has with those who are uninfected, the greater the value of R_0 and the faster the disease will spread. An increase in the population size or in the rate of transmission increases R_0, whereas an increase in parasite mortality or a decrease in transmission reduces the spread of disease in a population. Thus, a change that increases the value of R_0 tends to increase the proportion of hosts infected (prevalence) as well as the burden (incidence) of a disease. Usually, as the size of the host population increases, so do disease prevalence and incidence.

If the value of R_0 is larger than 1, the "seeds" of the infection (i.e., the transmission stages) will lead to an ever-expanding spread of the disease—an epidemic—but in time, as the pool of susceptible individuals is consumed (like fuel in a fire), the epidemic may eventually burn itself out, leaving the population to await a slow replenishment of new susceptible hosts (providing additional fuel) through birth or immigration. Then a new epidemic may be triggered by the introduction of a new parasite or mutation, or there may be a slow oscillation in the number of infections, eventually leading to a persistent low level of disease. However, if R_0 is less than 1, then each infection produces fewer than one transmission stage and the parasite cannot establish itself.

The SARS (severe acute respiratory syndrome) outbreak in 2003—which curtailed travel and commerce and led to losses on the order of hundreds of billions of dollars—was controlled by quarantine because early in the outbreak the R_0 value was calculated to be 3.0, meaning that a single person with an infectious case of SARS would infect about three others if control measures were not instituted. This value suggested that there was a low to moderate rate of transmissibility and that hospitalization would block the spread of SARS. The prediction was borne out: transmission rates fell as a result of reductions in population contact rates and improved hospital infection control, as well as more rapid hospitalization of suspected (but asymptomatic) individuals. By July 2003, the R_0 value was much smaller than 1. For a disease such as influenza, with its high R_0 value of 10, quarantining those who show symptoms would not be enough to bring the average number of new infections caused by each case to below 1: the level necessary for an epidemic to go into decline. Even though the R_0 value for the 1918 pandemic has recently been calculated to be lower than earlier estimates, i.e., R_0 is 1.8 to 3.0, quarantine would not have worked to stem its spread because (i) the influenza virus is actively being shed for the first 2 to 4 days after infection, (ii) infectivity is maximal early in the illness, and (iii) there are a large number of mild cases without symptoms that may be able to transmit the infection. Such conditions provide little time for isolating the sick and halting transmission. (Quarantine worked with the SARS epidemic because the virus is shed many days after infection and infectivity peaks after 2 weeks—conditions quite different from that of influenza.)

Since quarantine would not control an influenza pandemic, other measures, such as treatment and immunization, would have to be employed. Treatments involve antiviral drugs (zanamivir [marketed as Relenza] and oseltamivir [marketed as Tamiflu]) that block the synthesis of the N protein. Given that influenza viruses cannot move from cell to cell without N, such drugs, when administered early enough, eliminate the virus because complete virus cannot be released. Zanamivir, unlike oseltamivir, cannot be taken orally and must be inhaled. Both drugs are expensive and are usually in short supply. Other antiviral drugs (amantadine and rimantadine) target a protein that is required by the virus to release its RNA after entering the cells of the respiratory tract, and so the virus is unable to replicate. However, these drugs are not perfect, and the viruses quickly acquire resistance to them. (Of great concern is that the 2004 H5N1 strain is resistant to amantadine.) All of these drugs are less effective in preventing infection, but they may reduce the severity as well as the duration of symptoms.

Consequences

Another influenza pandemic is inevitable. Modern means of transportation—especially jet airplanes—ensure that the virus can be spread across the globe in a matter of hours or within a day in an infected traveler. Surveillance of birds may alert us to the possibility of an outbreak, and vaccination of birds and pigs may reduce the chances of viruses jumping into humans, but this will require abundant financial and logistical resources. Drugs may reduce the severity of illness, but neither surveillance nor culling can guarantee when, where, or how lethal the next pandemic will be. What is predictable is that it will seriously impact our lives: hospital facilities will be overwhelmed because medical personnel will also become sick; vaccine production will be slower because many of the personnel in pharmaceutical companies will be too ill to work; and reserves of vaccines and drugs will soon be depleted, leaving most people vulnerable to infection. There will be social and economic disruptions. A further complication is that many pharmaceutical companies are reluctant to manufacture vaccines because the risks involved with a lack of predictability, of consumer demand, as well as the absence of financial incentives may lead to vaccines becoming "money losers." Some of the remedies to ensure a reliable vaccine supply would be to allow collaborations between industry, academe, and the government and to provide economic incentives (including fair pricing, guaranteed purchase of unsold supplies, regulatory relief, tax incentives, and liability and intellectual property protection) to vaccine manufacturers. Even with such positive steps, scaling up the production of vaccines would not be easy. Although 300 million doses are made each year, billions of doses might be required if a pandemic arises. If protective vaccines do become available, there will be great social challenges: who should be vaccinated, at what age, and at what cost?

We need to know more about what makes an influenza virus transmissible to humans, why some cells are more susceptible to the virus and others are not, and how the virus interacts with the human immune system. If we can better define the environmental factors that allow the virus to spread, we might be able to limit the next epidemic. How can we prepare for a "future shock" such as that of the 1918 to 1920 pandemic? Stockpiling anti-infective drugs, promoting vaccine development, and increasing the methods for surveillance may all help blunt the effects, but unfortunately these measures cannot eliminate them.

11

AIDS: the 21st Century Plague

Imagine having this nightmare: you are infected with the human immunodeficiency virus (HIV). When you tested positive 4 years ago, you were in good health, but since that time you have begun experiencing memory loss and difficulty with your powers of concentration. The bouts of night sweats, persistent diarrhea, and weight loss have forced you to leave your job. When you look in the mirror, you see a body in decay—hollow-cheeked pallor and a severe wasting so that the bones of your rib cage protrude. Two years ago you had your first hospitalization for meningitis, an inflammation of the brain which caused such painful headaches that you could hardly stand being in the sunlight. Your blood count—especially the numbers of a specific kind of white cell (called a helper T cell)—was abnormally low: it was at 500. "Five hundred," you said to the doctor, "What does that mean? Is it bad?" The doctor reassured you that this would be taken care of by a prescription for a new cocktail of drugs. It has not worked, and now, when your helper T-cell count has declined to below 200, the telltale symptoms of AIDS have appeared: you are emaciated, your skin is covered with purple-black blotches, and your gums are badly swollen and covered with thrush, a whitish fungus. The bright hope for a magic bullet has faded. Your lungs are being ravaged by *Pneumocystis jiroveci* (previously named *P. carinii*) pneumonia, and breathing is becoming increasingly difficult. Today you were unable to climb a flight of stairs. More and more time is spent in bed thinking about only one thing: the prospect of death. You ask: Why haven't the drugs worked their magic? How long will I be able to afford treatment? What good is science if it understands HIV so well and yet it can't produce a vaccine? What has crippled my immune system so that it cannot cope with ordinary fungal and bacterial infections? Where did HIV come from? Will I suffer from isolation, discrimination, and humiliation? What is the government doing to cure me and help protect those who are uninfected?

Cause

The story of the discovery of HIV begins more than a century ago (1884) with the development of a porcelain filter by Charles Chamberland, who was working in the laboratory of Louis Pasteur. Chamberland's filter had very small pores, and it was possible by using this filter, as Chamberland wrote "to have one's own pure spring water at home, simply by passing the water through the filter and removing the microbes." In 1911, a sick chicken was brought to a young physician, Peyton Rous (1879 to 1970), who was working at the Rockefeller Institute in New York City. The chicken had a large and disgusting tumor in its breast muscle. Rous wondered what could cause such a tumor (called a sarcoma), and so he took the tumor tissue, ground it with sterile sand, suspended it in a salt solution, shook it, centrifuged it to remove the sand and large particles, and filtered it through a Chamberland filter. The sap (or filtrate) was used to inoculate chickens, several of which developed sarcomas a few weeks later. Rous examined the filtrate and the sarcoma under the light microscope and found that neither contained bacteria. He concluded he had discovered an infectious agent capable of causing tumors, that the infective agent was smaller than a bacterium, and that in all probability it was a virus. Rous failed to find similar virus-causing cancers in mice or humans and received no support for his beliefs from other scientists, so the sarcoma story was regarded as a biological curiosity. As a result, Rous turned his attention to other aspects of pathology. Ironically, just 4 years before his death at age 91, Rous' 1911 work was recognized when he received the 1966 Nobel Prize.

In the 1950s it became possible to grow a variety of cells in laboratory dishes (in vitro), and with these tissue cultures it became possible to study the effect of viruses on isolated living cells rather than in mice or chickens. At this time it was also found that there were two kinds of virus: those that contain DNA and those that contain RNA. Renato Dulbecco, who began to study DNA viruses in tissue culture cells, found that although some viruses caused tumors, on occasion virus could not be detected in the infected cell because its genetic material (DNA) had become inserted or integrated into the cell's chromosomes. In other words, the virus had been incorporated into the host cell genes and behaved as if it were a part of the cell's genetic apparatus. The virus DNA was now a part of the cell's heredity!

Dulbecco worked with DNA viruses, and this made it easier to visualize how the DNA of a virus could be integrated into the DNA of the host cell. However, Rous sarcoma virus (RSV) was an RNA-containing virus, so it was not obvious how the genetic information of RSV could become a part of the tumor cell's heredity. Howard Temin and David Baltimore provided the answer. Temin (1934 to 1994) attended Swarthmore College, majoring in biology, and went on to CalTech to do graduate work with Renato Dulbecco; his doctoral thesis was on RSV. Temin's experiments, carried out from 1960 to 1964 at the University of Wisconsin, convinced him that when the RSV nucleic acid was incorporated into the genetic material of the host cell, it acted as a "provirus." Under appropriate conditions, the provirus would trigger the cell to become cancerous.

A fellow graduate of Swarthmore College, Baltimore (born 1938) did his doctoral work at the Massachusetts Institute of Technology and at the Rockefeller Institute, where he studied the virus-specific enzymes of RSV. His first independent position was at the Salk Institute in La Jolla, Ca., with Renato Dulbecco, studying RSV in tissue culture. When he returned to the Massachusetts Institute of Technology as a faculty member, he continued to study the RSV enzymes. In 1970 Temin and Baltimore simultaneously showed that a specific enzyme, reverse transcriptase, in RSV was able to make a DNA copy from the viral RNA. They independently went on to show that the replication of RNA viruses involves the transfer of information from the viral DNA copy and that the viral DNA was integrated into the chromosomes of the transformed cancer cells. Later, other investigators were able to show that when purified DNA from the transformed cancer cell was introduced into normal cells, it triggered the production of new RNA tumor viruses. Clearly, RSV was a special kind of cancer-causing virus. For their discoveries, Dulbecco, Temin, and Baltimore shared the 1975 Nobel Prize for Physiology or Medicine.

This was the setting when AIDS appeared in 1981. Two laboratories—one in France headed by Luc Montagnier and one at the National Institutes of Health in the United States headed by Robert Gallo—identified a virus that was named HTLV III (human T-cell lymphotropic virus) by Gallo and LAV (lymphadenopathy virus) by Montagnier. The virus was found in tissues of patients with AIDS. Today both are recognized to be the same virus. The virus was renamed human immunodeficiency virus (HIV), and the disease complex it produced was called acquired immunodeficiency disease syndrome (AIDS).

Viruses are unable to replicate themselves without taking over the machinery of a living cell; in this sense, they are the ultimate parasites. The material containing the viral instructions for replication, i.e., its genes, may be composed of one of two kinds of nucleic acid, DNA or RNA, and the viral nucleic acid is packaged within a protein wrapper called the core; this in turn is encased in an outer virus coat or capsid, and the outermost layer is called the envelope. In HIV the genetic material is in the form of RNA, not DNA, so in order for this virus to use the machinery of the host cell (which can only copy from DNA) it must subvert the cell's machinery to copying viral RNA into DNA. As a consequence, in HIV the information flows from RNA to DNA to RNA to protein, and because the flow of information appears to be the reverse of what is typically found in cells (DNA to RNA to protein), these viruses (which include RSV) are called retroviruses.

A Crippled Immune System

HIV is able to cause AIDS because it can infect and destroy the white blood cells critical to the normal functioning of the immune system. The human body contains about a trillion white cells, and they are of five sorts. There are three kinds of granule-bearing white cells, called eosinophils, basophils, and neutrophils, and there are two kinds that lack granules in the cytoplasm, called lymphocytes and monocytes (macrophages). The lymphocytes and macrophages are produced in the bone marrow but are found in regional centers such as the spleen and lymph nodes as well as in the blood. Lymphocytes are divided into two different types called T and B lymphocytes. The B lymphocytes make antibody either on their own or by being activated by a T helper cell, called T4. The T4 lymphocytes are so named because they have on their surface a receptor molecule, CD4. Macrophages also have CD4 on their surface. Another kind of T lymphocyte, called a killer cell (also referred to as a T8 lymphocyte), is also activated by the T4 helper cell. T cells do not make antibody, but they are involved in what is referred to as cell-mediated immunity. The T cells communicate with one another by using soluble chemicals called chemokines that attract or activate other white cells, especially T and B lymphocytes and macrophages. To be activated, these white cells must have on their surface a receptor (analogous to a docking station) for the chemokine. Chemokine receptors also act as an entry cofactor that guides the viral capsid proteins into a shape that permits fusion to and entry of T cells and macrophages by HIV.

Now let us look more closely at the T lymphocytes—the cells that can be infected with HIV—and how virus and host cell interact with one another. The glycoproteins of the HIV capsid (which protrude through the envelope) resemble lollipops: the "stick" is called gp41, and the "candy ball" is gp120. The functions of these two viral proteins are to bind and anchor the HIV to the surface of the cell: CD4 on the surface of the T cell allows for the docking of gp120; once docked, the gp120 changes its shape so that it can bind to the chemokine receptor (called CCR5), and fusion and entry of HIV take place after binding.

One of the characteristics of an HIV infection is depletion of T4 cells. Other than direct lysis to release virus, it is not known precisely how the virus depletes T4 cells, but it may involve a depression in the ability to expand their numbers. In a healthy individual there are ~1,000 T4 cells/mm^3 of blood, and in an HIV-infected individual the number declines by about 40 to 80 T4 cells/year. When the count reaches 400 to 800 cells/mm^3, the first opportunistic infections appear. Opportunistic infections are those that under ordinary circumstances cause no disease but, in individuals with a weakened immune system, take advantage of the opportunity afforded them by the body's crippled defenses and cause clinical disease. The first opportunistic diseases are annoying infections of the skin and mucous membranes: thrush, involving painful sores in the mouth (and also sometimes on the vulva and prepuce), due to the fungus *Candida albicans*, and shingles, a viral disease of the peripheral nerves, due to herpes zoster virus (chickenpox virus), usually appear 1 to 3 months after infection. In addition, severe athlete's foot and white patches on the tongue, due to Epstein-Barr virus, can occur. Once these symptoms appear, the person is said to have AIDS-related complex (ARC). Such individuals may also have lymphadenopathy, excessive weight loss (10 to 15% of their body weight), profuse night sweats, fevers, persistent cough, and diarrhea. Some individuals may be infected but not show ARC for years, yet such individuals are infectious.

Once the T4 count drops below 200, the individual is said to have AIDS. At this T-cell level the AIDS-defining opportunistic infections are *Pneumocystis* pneumonia, cryptococcal meningitis, and toxoplasmosis (and these result in 50 to 75% of deaths). When the T4-cell count drops below 100 T4 cells/mm^3, several other opportunistic infections occur: histoplasmosis; infection with cytomegalovirus, a herpesvirus leading to blindness, lung damage, and digestive tract pathology; *Mycobacterium avium* (bird tuberculosis) infection; and chronic cold sores due to herpes simplex virus. Opportunistic diseases can be more than debilitating; together, they cause 90% of AIDS-related deaths. In addition, some other conditions may occur: encephalopathy leading to hallucinations and dementia, motor loss, and lymphomas. Few people doubt that HIV causes AIDS, yet a perplexing feature of this viral infection is why the disease progresses so rapidly in some individuals but not in others. For example, some people may be asymptomatic for up to 10 years whereas others develop AIDS symptoms within a year of initial infection.

Inherited Resistance

For HIV to infect a cell, it must first attach (see above). Since the attachment process involves linking to the CCR5 chemokine receptor, a person with a mutation in the receptor, such that it is absent from the cell surface, would not be susceptible to HIV infection because the virus would be unable to infect the T cell. Approximately 10% of Caucasians of western European descent have this chemokine receptor mutation called CCR5-Δ32, and the trait is virtually absent from African, Middle Eastern, Asian, and American Indian populations.

How and when did this mutation arise? It has been suggested that the high frequency of this variant in Europe arose through strong selection from the high fatality rate during epidemics of smallpox. Probably the mutation first occurred in northern Europe 1,000 to 1,200 years ago and then spread in a continuous gradient down to the south and the Mediterranean coast by Viking dispersal. The descendants who carry the Δ32 mutation suffer no illness; however, the absence of CCR5 prevents the smallpox virus from docking onto the white blood cell surface and so these cells can neither be invaded nor act as "ferry boats" to carry the virus to other locations in the body. In effect, the mutation protects the individual against the lethal effects of smallpox. In addition to CCR5 acting as a docking site for the smallpox virus, it can serve as a receptor for another virus: HIV. Thus, CCR5-Δ32 acts to protect against HIV, but to be completely effective it must be inherited in a double dose; one copy of the mutation delays the onset of AIDS because the invasion efficiency of HIV is reduced by about 50%, but it does not prevent it.

"Catching" HIV

HIV causes an infectious disease but not a very contagious one. It is not spread very easily via fomites (inanimate objects) since the virus is killed by high temperature, detergents, and 10% chlorine bleach. Neither is it a vector-borne virus spread by blood-sucking insects. HIV is found principally in the secretions and body fluids of infected individuals. However, the most infectious sources are blood, breast milk, vaginal secretions, and semen, where there are sufficient concentrations of virus to cause infection. Viruses are found both free and inside T lymphocytes. A tear or lesion in the skin or mucous membranes must occur for infected fluid to initiate an infection. Alternatively, infection may result from injection of infected blood into the body, organ transplants, or ingestion of infected breast milk; the virus can also cross the placenta.

The spread of HIV through the community can be thought of as a process of diffusion, where the motions of the individuals are random and movement occurs from a higher concentration to a lower one. Therefore, factors affecting the spread include the size of the population, the communal activities that serve to bring susceptible individuals in contact with infectious individuals, and the preventive measures used. Using the formula $R_0 = \beta Dc$, where β is the average probability of contact, D is the duration of infectiousness, and c is the number of contacts per unit time, it is possible to calculate the critical rate of sexual partner exchange that allows a disease such as AIDS to spread through a population, i.e., when R_0 is greater than 1. For HIV, with a duration of infectiousness of 0.5 years and a transmission probability of contact of 0.2, the partner exchange value is 10 new partners per year.

The value for R_0 increases with the transmission rate as well as the duration of the host's infectiousness. AIDS, a consequence of an increase in the virulence of HIV, is thought to have resulted from accelerated transmission rates of the virus due to changes in human sexual behavior: the increased number of sexual partners was so effective in spreading HIV that human survival became less important than survival of the parasite. There are three principal means of HIV transmission: (i) unprotected sexual intercourse that allows the virus to enter the body (transmission rates are lower between an infected female and an uninfected male because vaginal secretion is much less abundant than semen; however, HIV can be transmitted via oral sex, with cunnilingus being less efficient than fellatio); (ii) the sharing of HIV-contaminated blood and blood products; and

(iii) passage from an infected mother to her child as a result of small ruptures in the placenta and/or via breast milk.

From *Pan* to Pandemic

Where did AIDS come from? Where was HIV before the 1980s? Most people assume that the epidemic began in New York City and San Francisco among the homosexual community and then moved to Europe, but newer evidence points to the first appearance of AIDS in Europe among individuals who had never been to the Americas. One case was a Norwegian sailor who appears to have become infected some time before 1966. He developed lymphadenopathy with recurrent colds. He died in 1976 at age 29 with dementia and pneumonia. His wife and daughter also died of opportunistic infections. Blood serum samples taken from all three between 1971 and 1973 tested positive for HIV. The sailor had traveled to Cameroon in 1961 and 1962 and had been treated for gonorrhea during the trip; it is likely that he became infected at that time and subsequently passed the infection on to his wife and child. In another case, a sailor in Great Britain in 1959 had all of the telltale signs of HIV; his only trip outside of Britain was to Morocco. Although by 1962 there were 33 cases of opportunistic infections similar to those seen in AIDS patients, and 80% of these infections were found in men in their 30s, no one suspected that these European infections would portend an epidemic.

There are three types of HIV: HIV-1, discovered in 1983; HIV-2, discovered in 1985 and confined mostly to West Africa; and HIV-0, discovered in 1990 and found only in Cameroon and Gabon. There are two subtypes of HIV-1: HIV-1A and HIV-1B. HIV-1A and HIV-2 are both spread heterosexually, but HIV-1A is more virulent than HIV-2. HIV-1B dominates in Europe and the United States. HIV-like viruses occur in cows, lions, horses, sheep, goats, and simians (monkeys), but in most of these natural hosts they cause little or no immunodeficiency. The simian viruses are called SIV. Discoveries in the 1980s in chimpanzees (*Pan troglodytes*) in the Gabon rainforest showed a similarity between the virus of the chimpanzee (called SIVcpz) and HIV: serum taken from infected chimpanzees or serum produced against SIVcpz by injection reacted with HIV-1 and vice versa. SIVcpz, transmitted among chimps sexually with no ill effects, entered the human population, or so we believe, through forest people who engaged in the bush meat trade (i.e., hunters who used primitive butchery methods to dismember chimpanzees and other apes and sell or eat the meat, either raw or improperly cooked, hence coming into contact with chimpanzee body fluids such as saliva and blood). This species jump of the virus from chimpanzee to human probably occurred in the eastern part of the west equatorial rainforest, that is, in the eastern part of Cameroon, northern Congo, and the West African Republic. Once the virus had entered the human population, it was transmitted sexually. It has been speculated that over a period of 200 to 400 years it evolved into HIV-0 and HIV-1, with the latter becoming the much more virulent HIV-1A over time.

Where did HIV-1B come from? It probably came to Europe from Africa, and the most likely region is Cameroon, which was at one time German East Africa. In the 1880s the Europeans partitioned Africa into many colonies. The French had West Africa; the British settled in Sierra Leone, Ghana and the Cape, and Egypt; and Germany was active in Togo, Cameroon, and West Africa. European colonialism caused considerable turmoil in Africa, and it was not uncommon for the Europeans to have sexual relations with the native population. It is thought that the movements of the soldiers and sailors from West Africa to East Africa took the virus with them. By the end of World War I, when the German colonies were redistributed among the victors—the French took 80% of Cameroon, and Britain took 20%—HIV-1 had already been seeded around Lake Victoria, and in time it would slowly emerge as HIV-1A. HIV-1B was probably introduced into Europe in 1939 by the 300 Germans who returned to Germany from Cameroon via Danzig. These were part of a group of Germans who had remained in Cameroon after World War I, but by 1939, when war was declared and World War II began, they lost their land and were repatriated to Germany by ship.

During the 1980s in the Uganda-Tanzania region around Lake Victoria, a new disease was described—"slim disease"—characterized by diarrhea, weight loss, and fever. The disease appears to have come to Uganda from Tanzania via traders and soldiers. From 1900 to 1950 African HIV-1 appears to have been transmitted only fast enough to be maintained as an endemic disease but not fast enough to increase its virulence. However, in the late 1960s and 1970s, new opportunities arose through heterosexual activity of traders and soldiers and prostitution, and HIV-1A emerged. With its morbidity and mortality increased, as typified by the wasting "slim disease," it spread from its center around Lake Victoria beyond Uganda-Tanzania, moving along the truck routes. AIDS now became more and more widespread in Africa.

If the hunters of the African forest had HIV and it caused an AIDS-like disease, why did it go unnoticed for so many years? Probably because it remained isolated in a small population and European contact with these forest people did not take place until the 1890s. Further, within Africa a disease of this sort might not be noticed as being special and different from the wasting that occurred with malnutrition or infections such as malaria, yellow fever, and hookworm. Precisely when the virus spread to other human populations and diversified remains uncertain; however, it appears to coincide with a time when socioeconomic and political changes were occurring in Africa; i.e., during the 1950s and 1960s colonial rule in Africa ended, there were civil wars, vaccination programs involved the reuse of needles and syringes, large cities developed, there was a sexual revolution, and travel within Africa and between Africa and the rest of the world increased.

Several SIVs have been isolated from Japanese macaques, African green monkeys, African chimpanzees, and sooty mangabeys. The SIV from the sooty mangabey, called SIVsm, is most closely related to HIV-2. Indeed, antibodies to HIV-2 and SIVsm show these two viruses to be indistinguishable. Ten percent of West African prostitutes test positive for both SIVsm and HIV-2, but in the United States and Central Africa there is little reaction with SIVsm. HIV-2 is thought to have arisen in West Africa in the rainforests of Benin, and to this day it remains largely confined to West Africa; it spreads poorly by heterosexual contact. HIV-2 is less pathogenic than HIV-1, and individuals infected with HIV-2 are less at risk for developing AIDS. Where did the sooty mangabey get its SIV? Perhaps from large cats harboring feline immunodeficiency virus, which caused a benign disease. SIVsm is being tested as a candidate vaccine against AIDS.

Chemotherapy

One possible approach to controlling the transmission of an infectious agent is to treat it with drugs. Treatments for HIV infections were not available for several years after HIV was identified, but through the efforts of George Hitchings and Gertrude Elion, who directed an antiviral program at Burroughs-Wellcome, all that would change. Hitchings and Elion were chemists-pharmacologists who had produced active compounds against the smallpox virus. Some of these were sent to the Sloan-Kettering Institute in New York to be screened for anticancer activity as well. In 1947 they had synthesized an active compound, 2,6-diaminopurine, that resulted in remission of acute leukemia. This encouraged the search for other antitumor agents, and by 1948 they had several compounds—analogs of nucleic acid bases—that could interrupt the rampant replication of cancer cells. When the AIDS pandemic began, Burroughs-Wellcome already had on hand a wide array of antitumor compounds. One was azidothymidine (AZT) (developed in 1964), which soon became the first-line treatment for AIDS (in the late 1980s). For their work on antiviral drugs, Hitchings and Elion received the Nobel Prize for Physiology or Medicine in 1988.

Understanding the way HIV replicates has provided the basis for the rational design of drugs to block retrovirus replication. Although at present there is no drug that will cure an HIV infection, some drugs are effective in slowing the spread of symptoms and prolonging life. The most commonly used drug, AZT, delays the onset of AIDS by inhibiting viral multiplication. In the case of a rapidly multiplying virus where RNA is copied into DNA by reverse transcriptase, incorporation of the abnormal nucleic acid base AZT (instead of thymidine) jams the cell's copier and, in so doing, blocks the synthesis of new virus particles. Although AZT is also incorporated into host cell DNA, it is incorporated at a much lower level than in the virus and therefore is not very toxic. However, because AZT inhibits red blood cell formation in the bone marrow, individuals treated with AZT do suffer from nausea, weight loss, and anemia. Other drugs such as ddC (dideoxycytidine) or ddI (dideoxyinosine) also block the synthesis of viral nucleic acids and work in a similar fashion. These other drugs, called nucleoside analogs or reverse transcriptase inhibitors, and AZT are not cheap and must be taken frequently, usually daily.

A viral enzyme that cuts up viral proteins, called a protease, is encoded by one of the nine HIV genes. Viral proteins are synthesized much like a run-on sentence, and the protease cuts the long viral protein into shorter fragments, producing, in effect, separate words, so that they can be assembled into a complete virus particle. Some of the newer drugs being used for treatment of HIV target this viral protease and act to inhibit the cutting process; if this inhibition works, then short viral proteins cannot be produced and a complete virus cannot be assembled. Protease inhibitors (indanivir, saquinavir, and ritonavir) are expensive. They have been combined with reverse transcriptase inhibitors to block viral replication, and the combination can reduce virus production by 90 to 99%; however, this treatment too can be costly. Then there is triple therapy, also called HAART (for "highly active antiretroviral therapy"), which includes, in addition to a protease inhibitor and a nucleoside reverse transcriptase inhibitor such as AZT, a nonnucleoside reverse transcription inhibitor.

Because AIDS takes a long time to manifest itself, it can be difficult to assess the effectiveness of a new compound after only a short time. As with most viruses, the AIDS-causing virus is hard to kill without doing any damage to the host. Even with the combination therapy, drug-resistant viruses do emerge. Treatment by a drug can be a two-edged sword: it may benefit the individual, but if it does not reduce infectiousness, it might not significantly benefit the community. It has been calculated that the number of virus particles must be reduced to less than 50/µl of serum for the individual to lose the capacity to be infectious. (However, it has been shown that, even with successful reduction, a viral reservoir remains in the body and that if therapy is discontinued, the virus rebounds.) A further problem with AIDS is that one must also treat an assortment of opportunistic infections, which are the infections that actually kill the patient.

How does drug resistance come about, and why might it be prevented by a drug cocktail? Drug resistance is the result of natural selection; i.e., genotypes that are the best able to survive and reproduce pass on their genes to future generations and increase in frequency over time. Drug resistance develops this way. Let us assume that an average HIV patient has 1 billion virus particles and that 1 in 10,000 of these carries a mutation that allows that virus to evade the lethal effects of the anti-HIV drug. Once the patient is treated with that drug, only the mutant viruses survive, and these are the drug-resistant ones. The result is that there are now ~10,000 virus particles, and these can reproduce in the presence of the drug and increase their numbers, and now almost the entire viral population is resistant. However, let us assume that a second, and equally effective, drug is added along with the first drug to the billion viruses. Again, only 1 in 10,000 is resistant to this drug. When the drugs are used together, there is less than 1 resistant virus. By adding a third drug, the possibility for a surviving virus is reduced even further. Thus, HAART makes it possible to delay or prevent the emergence of drug resistance.

Failure To Control

Soon after the identification of HIV, a specific and sensitive test for antibodies to HIV was developed. Using such a diagnostic test, it has been possible to screen the sera of large numbers of individuals and to determine those who are infected. (Such individuals are called serum positive or seropositive.) This simple test also had immediate and profound effects for public health, since blood supplies in the United States and other countries could be screened for HIV. By 1985, through the use of screening, it was possible to ensure that the blood supply was HIV free, thereby preventing millions of potential transfusion-related infections. In addition, the antibody test could be used in epidemiological studies to determine the global scope and evolution of the disease. The availability of the antibody test has allowed for the identification of individuals before they show clinical signs of the disease and permits a more accurate description of the true clinical course of HIV infections. The Food and Drug Administration recently approved a rapid HIV antibody test that can be performed outside the laboratory and provide results in about 20 min. Those with positive results should have a confirmatory test; however, seronegative individuals should recognize that although they are presently without evidence of infection, in time they could become positive and therefore a repeat of the test would be necessary to ensure that they are HIV free.

Although antibodies to HIV can be found in more than 40 million people worldwide, these individuals continue to be infected with the virus and may transmit the disease to others. Indeed, each day there are more than 14,000 new infections. Clearly, although HIV elicits a strong antibody response, these antibodies neither neutralize the virus nor clear it from the body, and there is no protection against reinfection. This is because the antibodies to HIV are unable to reach the critical receptor sites and do not block the conformational changes in gp120 that are necessary for HIV attachment and entry. Furthermore, HIV is able to persist for long periods in nondividing T cells, so that even if an effective antibody were produced, its large size would not allow it to enter the infected cell and kill the virus. It has been estimated that if a person had a reservoir of 100,000 infected T cells, it would take 60 years to eradicate the infection. In addition to its ability to hide within resting T cells, HIV is able to cripple the immune system by depleting the helper T cells not only by outright destruction (through viral release) but also by decreasing their rate of production.

Since the body is unable to mount a protective immune response by infection, one might ask whether it is possible to stimulate the immune system by vaccination, as is done with other viral infections such as smallpox, measles, polio, and influenza. In the decades since the identification of HIV as the cause of AIDS, more than 35 vaccine candidates have been tested in 65 clinical trials in 10 countries, involving more than 10,000 volunteers at a cost of more than $650 million annually. The vaccines have consisted of a weakened (attenuated) virus or portions of the virus alone or attached to benign virus particles or naked HIV RNA, but none has proved effective. This is because of the high mutability of the virus, the difficulty in developing an effective cell-mediated immune response, and the inability of the immune system to be triggered to act on the lymphoid tissues of the gut, which is the primary site of HIV replication immediately after infection, as well as being the source of memory T cells that are necessary for boosting an immune response on reexposure. Were such vaccines to be developed, what would the consequences be of administering it to individuals already infected with HIV? How would vaccination of immunosuppressed HIV individuals affect their response to other vaccines?

In the absence of a vaccine and an effective cure by drug treatment, what can be done to avoid infection and reduce the spread of HIV? There are several possible tactics: (i) abstinence; (ii) maintenance of a monogamous sexual relationship with an uninfected individual; (iii) avoidance of risky behavior, especially for high-risk groups such as those with many sex partners and those engaging in anal intercourse; (iv) use of needle exchange programs to prevent the use of contaminated needles; (v) emphasis on the effectiveness of condoms and diaphragms; (vi) male circumcision; and (vii) acyclovir treatment for HSV-2 (herpes simplex virus).

Although early in the AIDS epidemic the virus could be spread via blood products (e.g., preparation of factor VIII for hemophiliacs) and whole blood used during transfusion, since 1985, as a result of screening the blood supply, blood and blood products are unlikely sources of transmission. Health care workers may be at higher risk, but only if body fluids of an infected individual make contact with the bloodstream of the recipient.

"Safer sex" describes practices that minimize the transfer of bodily secretions that contain the virus. Using a condom during oral, anal, or vaginal sex can reduce the transmission of the virus, but the condom must be used correctly. It needs to be placed on an erect penis, leaving room at the tip for the ejaculate. Withdrawal should occur before the penis becomes flaccid, and the condom should never be reused. Latex condoms are perishable and can deteriorate with time and heat. Water-based lubricants such as K-Y jelly, not a petroleum-based one such as Vaseline, should be used.

Some shun the use of condoms, contending that they have a high failure rate and so there is little benefit in using them to protect against AIDS.

The truth is that condoms are very effective in blocking HIV transmission, but they must be used consistently and correctly. Most of the statistics showing high failure rates relate to the use of condoms for birth control, and the failure rate relates to the number of births of infants conceived when their father was using condoms. However, these failures may be due to inconsistent or improper use of condoms and not to an inherent flaw in the condoms themselves. At least one study shows that most failures come from a small minority of people who did not, on closer examination, follow the proper condom use steps outlined above.

Those who avoid the use of condoms sometimes justify this reckless behavior by claiming that HIV is so small that it can pass through the holes in a condom. These individuals know little about HIV or the molecular characteristics of the latex rubber in a condom. First, most of the viruses are inside T4 cells, and these cells are larger than sperm, which do not pass through a condom (otherwise condoms would be useless for contraception). Even the free virus does not pass through the condom wall because it contains millions of atoms, much larger than even the molecules in water, which do not pass through the intact condom wall either. Finally, some people are concerned about condoms breaking during use. Condom failure is very rare; condoms are rigorously tested during manufacture, and where ripping does occur it is commonly the result of improper storage and use of petroleum-based lubricants. The best evidence for protection by condoms comes from a study of couples in Europe in which one member was HIV infected and the other was not. Among 123 couples that reported consistent condom use, none of the uninfected individuals became infected, whereas among 122 couples that used condoms inconsistently, 12 of the uninfected partners became infected. "Safer sex," including condom use, minimizes risk but does not eliminate it entirely. In a similar fashion, while driving a car we may manage to reduce the risk of accidental injury or death by wearing seat belts and driving carefully, but we cannot eliminate threats to our existence completely.

Control by Behavior

The power of changes in risk behavior to control the spread of AIDS has been clearly shown in a theoretical transmission model using data from the San Francisco Young Men's Health Study. The model addressed three questions. (i) If an HIV vaccine were available, what proportion of young homosexual men in the community would have to be vaccinated in order to eradicate HIV? (ii) How effective would such a vaccine have to be in order to ensure eradication? (iii) What effect would changes in sexual risk behavior have if a mass vaccination program were instituted? In this model, the value for R_0 (the number of secondary cases that result when one infectious person is introduced into an uninfected population) was estimated to be between 2 and 5 when using sexual risk behavior data such as the number of receptive anal sex partners and the use of condoms. (At the time of the study, the infection rate in young homosexual men in San Francisco was 18%.) The model predicted, with $R_0 = 2$, that 80% of those vaccinated would have to become immune to ensure eradication; i.e., susceptibility to infection would have to be reduced by 95% and immunity would have to persist with a half-life of 35 years. However, if $R_0 = 5$, the minimum efficacy of the vaccine would have to be 80% and coverage would have to be 100% to achieve HIV eradication. Since participants in the study indicated that they would not participate in a vaccine trial if it were only 50% effective, it is possible that a mass vaccination program could actually increase the severity of the epidemic. In the absence of a change in risk behavior, and with $R_0 = 5$, a vaccine that was 60% efficacious could not eradicate the epidemic; even with a vaccine of 80% efficacy, 100% coverage would be needed. However, if risk behavior were halved, eradication would be possible with a vaccine that was 60% efficacious. The theoretical model predicted that without a mass vaccination campaign and solely a reduction in risk behavior, HIV could be eradicated. For example, if $R_0 = 2$, the risk behavior levels would have to be decreased by 50%, and for $R_0 = 5$, the decrease would have to be 80%. This model calculates that even when a highly effective vaccine does become available (and to date none exists), coverage will have to be very high to achieve eradication, and it is extremely unlikely that HIV eradication could occur in San Francisco without significant reductions in risk behavior combined with an effective vaccine. This model predicts that eradication of HIV will require simultaneous deployment of efficacious prophylactic vaccines and behavioral interventions. Today, behavioral prevention remains the only effective way to stem the spread of HIV.

The Social Context

In the year 1347, infected rats and their fleas were stirred up by traders cutting caravan routes through central Asia, bringing bubonic plague to Europe. In the space of 4 years, plague in its most virulent form, the Black

Death, killed up to 30 million people. AIDS, like the Black Death, engenders powerful conflicts. What role should the government play in promoting and protecting the public health? How do we weigh individual rights against the good of the community? What limits should be set on identifying and separating the sick from society? What is the social responsibility of the individual? Despite our greater knowledge of how microorganisms can cause disease and death and our sophistication in understanding modes of transmission of infectious diseases, the social responses to epidemics, AIDS included, follow a time-honored pattern. Epidemics or plagues are the result of a complex interplay of biological and social factors, and despite the uniqueness of each disease and a very different social structure from that of the past, there is a tendency for history to repeat itself. In brief, people do not react kindly to an epidemic disease even if they understand its etiology. Indeed, it is possible to see parallels between medieval reactions to bubonic plague and some of the contemporary fears regarding AIDS. Some of this stems from a crisis of confidence in physicians, the medical establishment, and the government, as well as xenophobia.

A similar pattern of response is seen with an epidemic disease, whether its origins are known or not. At first there is fear and anxiety, followed by flight and denial. The next step may be a justification of the changes to be instituted: "the public must be protected at all costs" becomes the clarion call. For example, there may be requests to establish screening programs to identify carriers as well as the sick, and/or there may be a call for public health measures to remove and isolate (quarantine) infected individuals. A politics of disease may also develop: bureaucratic agencies are established, and these begin to control resources for public health measures, diagnosis, treatment, research, and vaccine development. Emergency measures may be instituted: civil rights may be abrogated, as might the rights to privacy and freedom of movement, and the right to confidentiality may be rescinded. Disease, even if we know its cause, manner of transmission, and possible controls, may act to buttress social divisions and focus the religious, political, and cultural biases of a society. As a consequence, even in a sophisticated society, a cry goes out: "find the source of infection." Because the groups with which AIDS is frequently associated within the United States are often those held in low esteem and discriminated against in housing, jobs, and everyday social contacts (i.e., those of a different sexual persuasion, promiscuous individuals, drug addicts, prostitutes), we may hear: "Let's get rid of the poor, the homosexuals, the drug users, the unwed mothers, the prostitutes," and so on, and so on. Then there may be a call from the public for voluntary or enforced quarantine of infected persons without clinical signs since they could be dangerous carriers. Experience, however, shows that sustained quarantine for large numbers of people is not very successful, and since AIDS is not spread through casual contact or droplet infection, quarantine would be of little value in protecting the community. Once we become aware that contagious diseases may produce specific patterns of social and political responses, we may be better able to deal realistically and effectively with severe outbreaks—epidemics. And as George Santayana cautioned, "Those who do not remember the past are forced to repeat it."

Consequences

The AIDS epidemic is not uniform across the globe. There is not one epidemic but many. Each of these epidemics has a different dynamic and course, varying from place to place and time to time. The differences can be subtle, but they are important enough to require different and sometimes novel strategies.

Since 1981, AIDS has killed more than half a million Americans, a total exceeding all American combat-related deaths in all wars fought in the 20th century. The increase in the number of individuals infected is staggering: in 1981 there were 12 cases in the United States, now there may be 400,000 cases. Each year, 40,000 American become infected. However, concern about HIV/AIDS is declining in the United States. Today only 17% consider AIDS the "most urgent problem in the nation." Some of this may be due to the dramatic decline in the number of AIDS-related deaths as a result of the increased availability of HAART. In addition, because the profile of the AIDS patient has gradually shifted from white middle-class homosexual men to poor black and Hispanic residents of the inner cities and the rural South, the general public finds the epidemic less alarming. To prevent an increase in the number of AIDS cases, there must be a greater knowledge of an individual's HIV status. Without this information, an infected individual cannot know whether he or she is a transmitter and will not seek treatment. It is estimated that of the 850,000 to 950,000 Americans who may be infected, one-quarter are unaware they carry HIV. Of further concern is that the number of men having sex with men still accounts for the largest number of AIDS cases and the number of infected persons in this group is increasing. This may be due to the belief that HIV/AIDS is no longer considered a fatal disease, to an increase in the use of recreational drugs such as methamphetamine, to the easier access to anonymous sex partners through the internet, and to "prevention fatigue"—a lack of interest in hearing the same prevention messages repeatedly.

AIDS in Africa bears little resemblance to the epidemic in North America and Europe. In Africa the disease has a Darwinian perversion: the fittest of society, not the weak, die. Adults in their prime perish, leaving behind the old and the young. Everyone who is sexually active is at risk. Babies are infected by their HIV-infected mothers. Most infected people are unsure how they acquired the disease, but with their immune system ruined it is a certainty that they will all die from opportunistic infections. The AIDS epidemic has been at its worst in sub-Saharan Africa, where the prevalence is 8% in the adult population (ages 15 to 49) and there are 3.8 million new infections each year. In seven countries, all in the southern cone, the adult prevalence rate is 20%; Botswana has the highest rate at 36%, followed by Swaziland, Zimbabwe, and Lesotho at 25%. The driving force in these countries is the ease of mobility that comes with an extensive transport system and work migration, especially of miners. Other parts of Africa also have high rates of HIV infections: Côte d'Ivoire is among the worst-affected countries in the world, and Nigeria has a prevalence of 15%. Since the average annual income in Africa is $350, the cost of triple therapy (currently $10,000 in the United States) is prohibitive. Treatment of all the HIV infections would require $2 billion per year. Without treatment, those with HIV will sicken and die from AIDS; without prevention, the infection rates will not be checked. The health care systems that are needed are not in place, effective leadership is absent, and the continent continues to suffer from political turmoil.

The epidemic sweeps relentlessly across Africa, and its path has shifted from those at highest risk to a generalized one and from one concentrated mainly in urban areas to rural ones as well. Transmission is mostly heterosexual. There are, however, more than 1 million HIV infections in children younger than 15, and these have been acquired from their mothers. The number of children orphaned by AIDS is now over 13 million. At the end of the 1990s, the World Bank calculated that Africa's share of world exports dropped by nearly 60%, to 1.5% (from 3.5% in 1970). This loss is equal to a drop of 21% of the region's total economic output and more than five times the $13 billion that Africa receives in annual aid.

In more personal terms, where subsistence agriculture predominates, livestock is sold to pay for funeral expenses, and orphaned and unskilled children are left to look after the livestock and cannot attend school. Communities are saturated with orphans, many of whom must fend for themselves. They sink into penury, suffer from malnutrition and psychic distress, drift into lives of crime and prostitution, and contribute to social unrest. The human losses wreck already fragile economies, break down civil societies, and lead to political instability. Because AIDS in Africa is generally met with apathy, the infection rates soar, stigma hardens, denial hastens death, and the gulf between knowledge and behavior grows larger.

In Eastern Europe the epidemic is confined mostly to injecting drug users (IDUs), and it continues to spread rapidly as these countries experience economic and social upheaval; from the IDUs the disease is spread to their partners and sex workers. Vietnam has a lower HIV prevalence than the United States, but, as with the eastern European countries, the problem is with IDUs. It is estimated that 130,000 HIV infections occur among the 80 million Vietnamese, and there has been little success in slowing the spread among the IDUs. These infections are now spilling over into China. In Myanmar (formerly Burma) and Thailand, the plague of AIDS is fueled by a mix of drug users and sex workers, yet in Thailand it has been possible to stem the tide by changes in behavior and condom campaigns. Adoption of a "consistent condom use" policy among brothel-based sex workers has reduced the incidence among this group from 28% in 1996 to 13% by 2002. Myanmar has one of the worst HIV problems in Asia, and the virus is spread by prostitution and IDUs. Migrant workers—gem miners and loggers—provide the conduit to the general population. A political dictatorship, a lack of clinics, severe poverty, and a reluctance on the part of wealthy countries to invest or offer assistance make the disease problem worse.

These few examples show how HIV contributes to social vulnerability, reduces life expectancy, limits productivity, and stifles economic growth. AIDS exacerbates and prolongs poverty and increases malnutrition. HIV infections demand a greater proportion of the already meager annual income of those living in Asia, Africa, and Eastern Europe, thereby reducing access to food and health care. AIDS has diminished the number of available teachers, and in this way it impacts the educational system. Globally, HIV is one of the five leading causes of death. The modern plague of AIDS will continue to rise in the coming years as a result of infections that have already occurred, and it will decimate the ranks of the young men and women who are in their most productive years. The power that a plague could wield was clearly demonstrated 700 years ago by the Black Death. Today, the modern plague of AIDS is a forceful reminder that the global impact of infectious disease is yet to be blunted.

Epilogue

Centuries of experience with a dozen diseases provide convincing evidence that although we may never eliminate disease from the face of the planet, we can be optimistic that "coming plagues" can be rendered vulnerable. Indeed, despite the 1993 pronouncement by an International Task Force for Disease Eradication that none of the 95 diseases they considered possible candidates for eradication in a generation had in fact been driven to extinction, there is hope. We now realize that disease control rather than eradication is attainable; however, many diseases—some largely forgotten in many parts of the world—still cause misery, drain resources, and can threaten the health of others. Why? Is it because we have not learned important lessons from our past heroic battles with disease that can be applied today? Is it that we still lack a proper vaccine or an effective therapy? Is there some other factor?

Some contend that for effective disease control, we need an available vaccine or a "magic bullet." True as this may be, we have also found—much to our disappointment—that something more is required. Armed with an easily administered oral vaccine, public health officials in 1982 committed themselves to eliminating polio from the world by the year 2000. Since then, 2 billion children have been vaccinated, 5 million children have been protected from paralysis and death, and the incidence has been reduced by 99%, yet eradication is far from certain. Indeed, the persistent worldwide incidence of polio convinces us that the availability of an effective vaccine is not enough. In some countries a major obstacle can be the amount of funding, but there is also a more pernicious limit: ourselves. Or, as the cartoon character Pogo said: We have met the enemy, and he is us! In 2003 Nigeria halted its polio vaccination campaign for a year after rumors circulated that the vaccine contained HIVs and that the program was nothing more than an American scheme to sterilize Muslim girls. As a result, in just a few years, 18 once polio-free countries were experiencing polio outbreaks traceable to Nigeria. As the number of cases continues to decline and an eradication program drags on for what seems a time without end, fatigue sets in among volunteers, donors, and the public. The trouble with contagious diseases is that even a few unvaccinated children can create a new pocket of disease, which can start another outbreak elsewhere in the world. Disease—like a small cinder—can spread downwind and start a fire (epidemic) virtually anywhere.

On 30 April 2006, the Sunday *New York Times* reported some of the other critical factors that may limit disease control. The polio vaccine must be kept chilled from the time it leaves the factory until it reaches a child's mouth, so freezers have to be provided to the vaccinators. Finding the resources for enough freezers can be difficult, but even when this is solved, having enough people to administer the vaccine can become a problem. In Nigeria, only women can enter a Muslim household if the husband is away, and women and children are better at persuading other mothers to vaccinate. However, many husbands have refused to let their wives leave home, and others wanted the jobs (and the pay of $3 per day) for themselves or they sent their daughters. "As a result teenage girls could be seen leaving with empty boxes, not understanding they were supposed to carry ice packs and 40 doses of vaccine. Others carried tally sheets they could not fill out because they could not read." The messages here are clear: education must be coupled with the provision of a vaccine; an effective public health infrastructure and persistent application are vital if a vaccine is to do its work.

Measles is another example of how a disease—even when we have learned our lessons well from this and other contagious illnesses—can elude elimination. The measles vaccine is cheap. A single dose costs about 15 cents, yet measles still kills 450,000 children worldwide each year. India, which has more measles deaths than any other country, has not made eliminating measles a national priority, and as a result next year 100,000 children will die needlessly. Effective protection against measles requires two vaccinations, and so there must be a second-dose campaign. A second-dose campaign, first used in Latin America in 1994, eliminated measles from the Western Hemisphere by 2002 and halved the death rate in Africa, but in India it is not the availability of monies or vaccine that is at fault but a failure in public health policy: the objective of public health programs in India is to vaccinate against one disease at a time rather than improving routine immunizations; as a result, measles has fallen to the sidelines as the polio eradication campaign takes the lion's share of the public health resources. Yet in neighboring Nepal, there is a story of success. Here "50,000 mothers, most of them illiterate, are the foot soldiers, in a campaign to slash the number of deaths from measles. The volunteers, organized by the government down to the ward level, deliver invitations to each household by hand, then follow up the night before with a reminder visit, and shout their message like town criers. Over the years this public health system has distributed 2-cent doses of vitamin A to children, an intervention that cuts child mortality to almost a quarter . . . They have handed out de-worming tablets that cost a penny and slash infant mortality. They have handed out packets of oral rehydration salts that cost only 6 cents to save the lives of children with diarrhea. And they have gone door to door to take children to the clinic for immunizations. And why do these women do it? The illiterate mothers say they are willing to do a job that pays no salary because it gives them a way to contribute and win respect." Clearly, where there is the will, there can be successful control.

The tragedy of disease control is that its promise is not being fulfilled. Control of a disease takes more than an understanding of the biology of the pathogen—it requires proper financing, national will, and strategies for winning the public's trust. Disease control also requires surveillance and projections of how the infectious agent will spread; then, too, there must be a consistent and equitable application of control measures by a robust health care system. In some instances, reaping the public health benefits derived from a scientific understanding of disease will necessitate institution of the most difficult of all interventions: a change in human behavior. We have learned much of value from past encounters with illness, but there will be no quick fix. Yet, with a desire for improved health for all, coupled with a sensible implementation strategy, the obstacles to disease control can be mitigated. A reduced burden of disease is not only reasonable but also attainable.

Notes

In writing this book, primary and secondary literature sources and especially *The Power of Plagues* (ASM Press, 2006) have been relied upon. These chapter notes consist of references that are keyed to page numbers in the text.

Chapter 1. The Legacy of Disease: Porphyria and Hemophilia

Page 1. L. K. Altman and T. S. Purdum, JFK file, hidden illness, pain and pills, *New York Times* 17 Nov. 2002.

Page 1–3. *Porphyria—a Royal Malady*; articles published in or commissioned by the British Medical Association (London: British Medical Association, 1968); I. Macalpine and R. Hunter, *George III and the Mad Business* (New York: Pantheon, 1970).

Page 3–4. V. Vadakan, Porphyria, curse of royalty, *Hosp. Pract.* **22** (1987):107–114.

Page 3–4. M. R. Moore, Biochemistry of porphyria, *Int. J. Biochem.* **25** (1993): 1353–1368; H. H. Billett, Porphyrias: inborn errors in heme production, *Hosp. Pract.* **23** (1988):41–60.

Page 5. A. Bennett, *The Madness of George III*. (London: Faber and Faber, 1992), 18–21, 25, 31, 56.

Page 6–7. J. Rohl, M. Warren, and D. Hunt, *Purple Secret*: *Genes, Madness and the Royal Houses of Europe* (New York: Bantam, 1998); I. Macalpine and R. Hunter, *George III and the Mad Business* (New York: Pantheon, 1970); F. Cartwright and M. Biddis, *Disease and History* (London: Sutton, 1972), p. 172–173.

Page 7. T. M. Cox, N. Jack, S. Lofthouse, J. Watting, J. Haines, and M. J. Warren. King George III and porphyria: an elemental hypothesis and investigation. *Lancet* **366** (2005):332–335.

Page 7–8. I. Macalpine and R. Hunter, Porphyria and King George III, *Sci. Am.* **221** (1969):38–43.

Page 8–9. R. F. Stevens, The history of hemophilia in the royal families of Europe, *Br. J. Haematol.* **25** (1999):25–32.

Page 8–9. P. Mannuci and E. Tuddenham, The haemophiliacs—from royal genes to gene therapy, *N. Engl. J. Med.* **344** (2001):1773–1779.

Page 10–13. J. P. Gelardi, *Born To Rule* (New York: St. Martin's Press, 2005).

Page 10–13. F. Cartwright, *Disease and History* (New York: Dorset, 1972), ch. 7, p. 167–196; for another perspective, see J. Kendrick, Russia's imperial blood: Was Rasputin not the healer of legend, *Am. J. Hematol.* **77** (2004):92–102; R. F. Stevens, The history of haemophilia in the royal families of Europe, *Br. J. Haematol.* **105** (1999):25–32.

Page 12–13. K. Rose, *King George V* (New York: Knopf, 1983).

Page 13–17. D. M. Potts and W. T. W. Potts, *Queen Victoria's Gene: Haemophilia and the Royal Family* (New York: Sutton, 1999); J. M. Packard, *Victoria's Daughters* (London: St. Martin's Press, 1998).

Page 16–17. P. Ziegler, *King Edward VIII* (New York: Knopf, 1990).

Chapter 2. The Irish Potato Blight

Page 19–20. http://ballinagree.freeservers.com/sumsorrow.html.

Page 20–22. R. English, *History of Ireland* (New York: Gill and Macmillan, 1991).

Page 22. T. Cahill, *How the Irish Saved Civilization* (New York: Random House, 1996).

Page 23–29. G. L. Schumann, *Plant Diseases: Their Biological and Social Impact* (St. Paul, Minn.: American Phytopathological Society Press, 1991), 1–20; G. L. Comfort and E. R. Sprott, *Famine on the Wind* (New York: Rand McNally, 1967), 70–89; J. S. Donnelly, *The Great Irish Potato Famine* (Phoenix Mill: Sutton, 2001); N. Kissane, *The Irish Potato Famine: a Documentary History* (Dublin: National Library of Ireland, 1995); E. C. Large, *The Advance of the Fungi* (St. Paul, Minn.: American Phytopathological Society, 2003); D. C. Daly, The leaf that launched a thousand ships, *Nat. Hist.* **105** (1996):24–32; C. Kinealy, How politics fed the famine, *Nat. Hist.* **105** (1996):33–35; C. Woodham-Smith, *The Great Hunger* (New York: Harper and Row, 1962).

Page 23–24, 30–31. H. Hobhouse, *Seeds of Change: Five Plants That Transformed Mankind* (New York: Harper and Row, 1987), 191–232.

Page 32. G. Garelik, Taking the bite out of potato blight, *Science* **298** (2002): 1702–1704.

Chapter 3. Cholera

Page 33. J. Franklin and J. Sutherland, *Guinea Pig Doctors: the Drama of Medical Research through Self-Experimentation* (New York: Morrow, 1984), 140; W. McNeill, *Plagues and Peoples* (New York: Anchor, 1976), 231.

Page 33–34. I. Sherman, *The Power of Plagues* (Washington, D.C.: ASM Press, 2006), 167; S. LaFranniere, In oil-rich Angola, cholera preys upon poorest, *New York Times*, 16 June 2006.

Page 34–35. P. Johansen, H. Brody, N. Paneth, S. Rachman, and M. Rip, *Cholera, Chloroform and the Science of Medicine: a Life of John Snow* (New York: Oxford University Press, 2003); N. Longmate, *King Cholera: The Biography of a Disease* (London: Hamish Hamilton, 1966); S. Johnson, *The Ghost Map: the Story of London's Most Terrifying Epidemic and How It Changed Science, Cities and the Modern World* (New York: Riverhead, 2002).

Page 35. P. Pharoah letter. *Lancet* **363** (2004):1552.

Page 35–37. Sherman, *The Power of Plagues*, 160–161.

Page 37–38. G. N. Grob, *The Deadly Truth* (Cambridge, Mass.: Harvard University Press, 2002), 104–107.

Page 38. C. E. Rosenberg, *Explaining Epidemics and Other Studies in the History of Medicine* (Cambridge, Mass.: Harvard University Press, 1992), 114–121.

Page 39–40. F. Cartwright, *A Social History of Medicine* (London: Longman, 1977), 102–110.
Page 40. Sherman, *The Power of Plagues*, 128.
Page 40–43. H. Markell, *Quarantine!* (Baltimore: Johns Hopkins University Press, 1997).
Page 43–47. C. J. Gill and G. C. Gill. Nightingale in Scutari: her legacy re-examined, *Clin. Infect. Dis.* **40** (2005):1799–1805; L. Strachey, *Eminent Victorians* (London: Continuum, 2002); Florence Nightingale obituary, *The London Times* 15 August 1910; E. Huxley, *Florence Nightingale* (London: Weidenfeld and Nicholson, 1975).
Page 47–49. R. L. Guerrant, B. Carneiro-Filho, and R. E. Dillingham, Cholera, diarrhea and oral rehydration therapy: triumph and indictment, *Clin. Infect. Dis.* **37** (2003):398–505.

Chapter 4. Smallpox: the Speckled Monster

Page 50–51. R. Preston, *The Demon in the Freezer* (New York: Random House, 2002).
Page 50–52. T. O'Toole, Smallpox: an attack scenario, *Emerg. Infect. Dis.* **5** (1999):540–546.
Page 53–55. J. Eyler, Smallpox in history: the birth, death and impact of a dread disease, *J. Lab. Clin. Med.* **142** (2003):216–220; D. Hopkins, *Smallpox* (London: Churchill, 1962); D. Hopkins, *Princes and Peasants* (Chicago: University of Chicago Press, 1983); Sherman, *The Power of Plagues*, 191–195.
Page 55–56. M. Oldstone, *Viruses, Plagues and History* (New York: Oxford University Press, 1998).
Page 57–66. C. Mims, *The War within Us. Everyman's Guide to Infection and Immunity* (San Diego: Academic Press, 2000); Sherman, *The Power of Plagues*, 199–206, 211–228.

Chapter 5. Bubonic Plague

Page 68. G. Boccaccio, *The Decameron*, quoted in P. Ziegler, *The Black Death* (New York: Harper, 1969), 46.
Page 69. D. Herlihy, *The Black Death and the Transformation of the West* (Cambridge: Harvard University Press, 1997).
Page 69–73. Sherman, *The Power of Plagues*, 72–83, 128–129.
Page 74. McNeill, *Plagues and Peoples*, 108.
Page 75–76. C. McEvedy, The bubonic plague, *Sci. Am.* **258** (1988):118–123; M. Drancourt and D. Raoult, Molecular insights into the history of plague, *Microbes Infect.* **4** (2002):105–109; R. D. Perry and J. D. Featherstone, *Yersinia pestis*: etiologic agent of plague, *Clin. Microbiol. Rev.* **10** (1997):35–66.
Page 76. C. Cunha and B. Cunha, Impact of plague on history, *Infect. Dis. Clin. North Am.* **20** (2006):253–272.
Page 77. R. Gani and S. Leach, Epidemiologic determinants for modeling pneumonic plague outbreaks, *Emerg. Infect. Dis.* **10** (2004):608–614.
Page 77–78. E. Caniel, Plague. *Encyclopedia Microbiol.* **3** (2000):654–661; D. Zhou, Y. Han, Y. Sang, P. Huang, and R. Young, Comparative and evolutionary genomics of *Yersinia*, *Microbes Infect.* **6** (2004):1226–1234; N. C. Stenseth et al., Plague dynamics are driven by climate variation, *Proc. Natl. Acad. Sci. USA* **103** (2006):13110–13115; M. Achtman et al., *Yersinia pestis*, the cause of plague, is a recently emerged clone of *Yersinia pseudotuberculosis*, *Proc. Natl. Acad. Sci. USA* **96** (1999):14043–14048.
Page 79. M. Wheelis, Biological warfare at the 1346 siege of Caffa, *Emerg. Infect. Dis.* **8** (2002):971–975.
Page 79. T. Inglesby et al., Plague as a biological weapon, *JAMA* **283** (200):2281–2289.

Page 79–80. T. Inglesby, R. Grossman, and T. O' Toole, A plague on your city: observations from TOPOFF, *Clin. Infect. Dis.* **32** (2001):436–445.
Page 81. Plague—worldwide incidence (Figure 8). *Int. J. Health Geogr.* **4** (2005):10–21; S. A. Berger, GIDEON: a comprehensive web-based resource for geographic medicine, http://www.cdc.gov/ncidod/dvbid/plague/index.htm.
Page 82. W. Orent, *Plague: The Mysterious Past and Terrifying Future of the World's Most Dangerous Disease* (New York: Free Press, 2004).

Chapter 6. Syphilis: the Great Pox

Page 83. Franklin and Sutherland, *Guinea Pig Doctors,* 25–27.
Page 84. E. Tramont, The impact of syphilis on humankind, *Infect. Dis. Clin. North Am.* **18** (2004):101–110; A. Singh and B. Romanowski, Syphilis: review with emphasis on clinical, epidemiological and some biologic features, *Clin. Microbiol. Rev.* **12** (1999): 187–209.
Page 85. G. Anta, S. Lukehart, and A. Meheus, The endemic treponematoses, *Microbes Infect.* **4** (2002):83–94.
Page 85–86. C. Fraser et al., Complete genome sequence of *Treponema pallidum,* the syphilis spirochete, *Science* **281** (1998):375–388.
Page 86–87. C. Meyer, C. Jung, T. Kohl, A. Poenicke, A. Poppe, and K. Alt, Syphilis 2001—a palaeopathological reappraisal, *Homo* **53** (2002):39–58.
Page 87–89. Sherman, *The Power of Plagues,* 265–266.
Page 89–90. E. Tramont, Syphilis in adults: from Christopher Columbus to Sir Alexander Fleming, *Clin. Infect. Dis.* **21** (1995):1361–1371.
Page 90–91. T. G. Benedek and J. Erlen, The scientific environment of the Tuskegee study of syphilis, 1920–1960, *Perspect. Biol. Med.* **43** (1999):1–30.
Page 91–93. Sherman, *The Power of Plagues,* 245–251, 258–259.
Page 96. R. Hare, *The Birth of Penicillin and the Disarming of Microbes* (London: Allen and Unwin, 1970); E. Lax, *The Mould in Dr. Florey's Coat: the Remarkable True Story of Penicillin* (New York: Little, Brown, 2004).
Page 97. C. Amabile-Cuevas, M. Cardenas-Garcia, and M. Ludgar, Antibiotic resistance, *Am. Sci.* **83** (1995):320–329; S. Levy and B. Marshall, Antibacterial resistance worldwide: causes, challenges and responses. *Nat. Med.* **10** (2004): S122–S127.
Page 98–101. Sherman, *The Power of Plagues,* 268–270; http://www.who.int/docstore/hiv/GRSTI/index.htm; CDC, Primary and secondary syphilis—United States, 2003–2004. *Morb. Mortal. Wkly. Rep.* **55** (2006):269–273; L. Ferguson and J. Vanada, Syphilis: an old enemy still lurks, *J. Am. Acad. Nurse Pract.* **18** (2006):49–55.
Page 101–103. A. Rompalo, Can syphilis be eradicated from the world? *Curr. Opin. Infect. Dis.* **14** (2001):41–44.

Chapter 7. Tuberculosis: the People's Plague

Page 104. L. Hutchinson and M. Hutchinson, *Opera, Desire and Death* (Lincoln: University of Nebraska Press, 1996); S. Sontag, *Illness as Metaphor and AIDS and Its Metaphors* (New York: Anchor, 1989).
Page 105–109. R. Dubos and J. Dubos, *The White Plague* (Boston: Little, Brown, 1952); T. Dormandy, *The White Death: a History of Tuberculosis* (London: Hambledon, 1999); T. M. Daniel, *Captain of Death: the Story of Tuberculosis* (Rochester: University of Rochester Press, 1997); F. Haas and S. Haas, The origins of *Mycobacterium tuberculosis* and the notion of its contagiousness, *in* W. Rom and S. Garay (ed.), *Tuberculosis* (Boston: Little Brown, 1996), 3–34.

Page 106–107. S. Grzbowski and E. Allen, History and importance of scrofula, *Lancet* **346** (1995):1472–1474.

Page 109. D. Morens, At the deathbed of consumptive art, *Emerg. Infect. Dis.* **8** (2002):1353–1358.

Page 110. Ancient tuberculosis identified? *Science* **286** (1999):1071; M. Caldwell, *The Last Crusade: the War on Consumption 1862–1954* (New York: Atheneum, 1988); T. Garnier et al., The complete genome sequence of *Mycobacterium bovis*, *Proc. Natl. Acad. Sci. USA* **100** (2003):7877–7882; S. T. Cole et al., Deciphering the biology of *Mycobacterium tuberculosis* from the complete genome sequence. *Nature* **393** (1998):537–544.

Page 111. G. N. Grob, *The Deadly Truth: a History of Disease in America* (Cambridge, Mass.: Harvard University Press, 2002).

Page 111–112. A. M. Kraut, Plagues and prejudice, *in* D. Rosner (ed.), *Hives of Sickness* (Rutgers: Rutgers University Press, 1995), 65–90.

Page 112–120. Sherman, *The Power of Plagues*, 286–293; D. E. Hammerschmidt, Bovine tuberculosis: still a world health problem, *J. Lab. Clin. Med.* **141** (2003):359; S. Blower et al., The intrinsic transmission dynamics of tuberculosis epidemics, *Nat. Med.* **1** (1995):815–821.

Page 121–124. F. Ryan, *The Forgotten Plague: How the Battle against Tuberculosis Was Won—and Lost* (Boston: Little, Brown, 1993).

Page 124–126. P. Small and P. Fujiwara, Management of tuberculosis in the United States, *N. Engl. J. Med.* **345** (2001):189–200; T. Frieden et al., Tuberculosis, *Lancet* **362** (2003):887–897; Sherman, *The Power of Plagues*, 297–299.

Page 125. *Treatment of Latent TB Infection—2003*, City and County of San Francisco Department of Health, TB Control Section, 4; http://www.cdc.gov/nchstp/tb/pubs/tbfactsheets/250101.htm.

Page 126–127. T. Doherty, Progress and hindrances in tuberculosis vaccine development, *Lancet* **367** (2006):947–949.

Page 127–129. C. Dye, Global epidemiology of tuberculosis, *Lancet* **367** (2006):938–940; E. Corbett et al., Tuberculosis in sub-Saharan Africa: opportunities, challenges and change in the era of antiretroviral treatment, *Lancet* **367** (2006):926–937; S. Sharma and J. Liu, Progress of DOTS in global transmission control, *Lancet* **367** (2006):951–954; CDC, Trends in tuberculosis—United States, 2005, *Morb. Mortal. Wkly. Rep.* **55** (2006): 305–308; M. Gandy and A. Zumla (ed.), *The Return of the White Plague* (London: Verso, 2003).

Chapter 8. Malaria

Page 130–132. *A Fictional Account Based on the Writings of Ronald Ross, Memoirs* (London: John Murray, 1923).

Page 132–134. I. W. Sherman. A brief history of malaria and the discovery of the parasite life cycle, *in* I. W. Sherman (ed.), *Malaria: Parasite Biology, Pathogenesis, and Protection* (Washington, D.C.: ASM Press, 1998), 3–10.

Page 135. R. Kupuscinsk, *Shadow of the Sun* (New York: Vantage, 2002).

Page 136. Sherman, *The Power of Plagues*, 145–146, 156.

Page 136–138. G. Harrison, *Mosquitoes, Malaria and Man: a History of the Hostilities since 1880* (New York: Dutton, 1978); W. Bynum and C. Ovary, *The Beast in the Mosquito: the Correspondence of Ronald Ross and Patrick Manson* (Amsterdam: Editions Rudolphi, 1998).

Page 137. P. deKruif, *Microbe Hunters* (San Diego: Harcourt Brace, 1926), 256–285.

Page 138. R. Ross, *Memoirs* (London: John Murray, 1923).

Page 138–141. B. Greenwood et al., Malaria. *Lancet* **365** (2005):1487–1498.

Chapter 9. Yellow Fever: the Saffron Scourge

Page 143. deKruif. *Microbe Hunters,* 286–287.

Page 144. Franklin and Sutherland, *Guinea Pig Doctors,* 205–206.

Page 144–145, 147–148. M. Oldstone, *Viruses, Plagues and History* (New York: Oxford University Press, 1998), 45–72; Sherman, *The Power of Plagues,* 340–342, 344–345.

Page 146–147. E. T. Savitt and J. Young (ed.), *Disease and Distinctiveness in the American South* (Knoxville: University of Tennessee Press, 1988).

Page 149–153. Quoted in Oldstone, *Viruses, Plagues and History,* 47; L. K. Altman, *Who Goes First?* (Berkeley: University of California Press, 1998),129–158; Franklin and Sutherland, *Guinea Pig Doctors,* 183–226; deKruif, *Microbe Hunters,* 286–307.

Page 155–156. T. P. Monath, Yellow fever: an update. *Lancet Infect. Dis.* **1** (2001): 11–20; M. Theiler, *The Development of Vaccines against Yellow Fever,* Nobel Lecture, 11 December 1951.

Page 156. C. Soares, Turning yellow: making the yellow fever vaccine fight other germs, *Sci. Am.* Apr. 2006, 22–23.

Chapter 10. The Great Influenza

Page 158–159, 165–167. J. M. Barry, *The Great Influenza: the Epic Story of the 1918 Pandemic* (New York: Viking, 2004).

Page 159–160, 167. Oldstone, *Viruses, Plagues and History;* M. Specter, Nature's bioterrorist. *New Yorker,* 28 Feb. 2005, 51–61.

Page 160–163, 172–173. R. Webster and E. Walker, Influenza, *Am. Sci.* **91** (2003): 122–129; K. Nicholson, J. Wood, and M. Zambon, Influenza, *Lancet* **362** (2003): 1733–1745; P. Palese, Influenza: old and new threats, *Nat. Med.* **10** (2004):S82–S87; W. Laver, N. Biscofberger, and R. Webster, Disarming flu viruses, *Sci. Am.* Jan. 1999, 78–87.

Page 163–164. E. D. Kilbourne, Influenza pandemics of the 20th century, *Emerg. Infect. Dis.* **12** (2006):9–13; G. Neumann and Y. Kawaoka, Host range restriction and pathogenicity in the context of the influenza pandemic, *Emerg. Infect. Dis.* **12** (2006):881–886.

Page 164–165. J. C. Obenauer et al., Large-scale analysis of avian influenza isolates, *Science* **311** (2006):1576–1580.

Page 164–165. C. S. Smith, French farmers shudder as flu keeps chickens from ranging free, *New York Times* 1 Mar. 2006; J. Taubenberger and D. Morens, 1918 influenza: the mother of all pandemics, *Emerg. Infect. Dis.* **12** (2006):15–22.

Page 168–169, 170–171. Sherman, *The Power of Plagues,* 17–18, 398–399.

Page 169–170. D. M. Bell and the WHO Working Group, Nonpharmaceutical interventions for pandemic influenza, international measures, *Emerg. Infect. Dis.* **12** (2006):81–87.

Page 171–172. I. Longini et al., Containing pandemic influenza at the source, *Science* **309** (2005):1083–1087; D. M. Bell and the WHO Working Group, Nonpharmaceutical interventions for pandemic influenza, national and community measures, *Emerg. Infect. Dis.* **12** (2006):88–94; R. Webster, M. Peiris, H. Chen, and Y. Guen, H5N1 outbreaks and enzootic influenza, *Emerg. Infect. Dis.* **12** (2006):3–13.

Chapter 11. AIDS: the 21st Century Plague

Page 175–179, 181–185, 187–191. Sherman, *The Power of Plagues,* 92–105, 112–115; A. Galvani and M. Slatkin, Evaluating plague and smallpox as historical selective pressures for the CCR-δ32 HIV-resistance allele, *Proc. Natl. Acad. Sci. USA* **101** (2003):15276–15279.

Page 186–187. W. Koff et al., HIV vaccine design: insights from the live attenuated SIV vaccines, *Nat. Immunol.* **7** (2006):19–23; P. Spearman, Current progress in the development of HIV vaccines, *Curr. Pharm. Des.* **12** (2006):1–21; M. Girard, S. D. Osmanov, and M. Kiery, A review of vaccine research and development: the human immunodeficiency virus (HIV), *Vaccine* **24** (2006):4062–4081.

Epilogue

Page 195–196. C. W. Dugger and D. G. Mc Neill, Jr., Rumor, fear and fatigue hinder final push to end polio, *New York Times* 20 Mar. 2006.

Page 196–197. C. Dugger, Mothers of Nepal vanquish a killer of children, *New York Times* 30 Apr. 2006.

Index